Why Water Plants Don't Drown

Survival Strategies of Aquatic & Wetland Plants

Why Water Plants Don't Drown

Survival Strategies of Aquatic & Wetland Plants

By Victoria I. Sullivan

Illustrations by Susan E. Elliott

Pinyon Publishing
MONTROSE, CO

Also by Victoria I. Sullivan

Scientific Papers in
The Flora of North America, American Journal of Botany, Protoplasma, Planta, Plant Systematics and Evolution, Plant Species Biology, Sida, Rhodora, Canadian Journal of Botany, Brittonia, Interdisciplinary Humanities, and *National Association for Humanities Education publication*

Fiction
Adoption (2010)

Memoir
Granny's Letters: A Georgia Wiregrass Pioneer Woman's Tragedy (2008)

Also by Susan E. Elliott

Scientific Papers in
Environmental Entomology, American Journal of Botany, Ecological Entomology, and *Madroño*

Fiction
Ophelia's Ghost (2008)

Non-Fiction
Remembering the Parables (2010)
Making the Most of WriteItNow 4 (2010)

Illustrations
Open the Gates: Poems for Young Readers by Dabney Stuart (2010)
Spilled Milk: Haiku Destinies by Gary Hotham (2010)

Cover Painting Copyright © 2012 by Susan E. Elliott

Illustrations Copyright © 2012 by Susan E. Elliott

Photograph of Victoria I. Sullivan Copyright © 2012 by Diane M. Moore

Photograph of Susan E. Elliott Copyright © 2012 by Gary L. Entsminger

First Edition: September 2012

Pinyon Publishing
23847 V66 Trail, Montrose, CO 81403
www.pinyon-publishing.com

Library of Congress Control Number: 2012948718
ISBN: 978-1-936671-10-6

ACKNOWLEDGMENTS

The people influential to my career as a botanist, wetland plant enthusiast, and writer of this book are more numerous than I can acknowledge, but I will name a few who stand out in my mind. Robert K. Godfrey, Florida State University professor, author of many books, and co-author with Jean W. Wooten of the excellent two-volume work on the aquatic and wetland plants in the southeastern United States set a high bar for plant exploration, collecting, and editing for his students, of which I was one, and inspired my early interest in aquatic and wetlands plants. I searched innumerable ditches, margins of lakes and ponds, and riverbanks throughout the southeast with Jean W. Wooten, a fellow graduate student at Florida State University, for plants in the genus *Sagittaria* (arrowheads) for her research, and moist pine flatwoods for plants in the genus *Aletris* for my research. Accompanying us were her daughter Beth and dog Katy, a water loving English Springer Spaniel. I also acknowledge the many students enrolled in my Aquatic and Wetland Plants classes at the University of Southwestern Louisiana (now University of Louisiana at Lafayette) whom I coaxed into unknown depths of water to grab aquatic plants over a 20-year span of teaching. They challenged me to continue learning and to inspire their interest in wetlands plants. I will always appreciate the encouragement to continue a graduate study at Florida State University by Frank C. Craighead, Sr. whom I had the pleasure of accompanying on field trips while working as a naturalist in Everglades National Park.

I acknowledge Gary Entsminger, a wonderful editor and publisher at Pinyon Publishing for seeing the potential in this book and for encouraging many profitable rewrites; and Susan Elliott, a consummate artist and botanist, who illustrated this book, never tiring of challenging me to be more clear and discerning, and without whose work this book would have been drab indeed.

Diane M. Moore, a prolific poet and writer par excellence, never ceases to challenge me to write and to write well. She offers her undying support, friendship, and affection, absolute necessities for my life and work.

To Robert K. Godfrey

CONTENTS

INTRODUCTION

Headlines:
FLOODS DESTROY VEGETABLE CROPS IN GEORGIA!
CORN CROPS DEVASTATED BY FLOODS IN THE MIDWEST!

Flooding is a serious problem for plants. But why do crops die when they're flooded for a prolonged period? Strange as it may seem, when plants are flooded, they die from lack of *water*. Specifically, their root hairs die from lack of oxygen (they suffocate), and subsequently, the rest of the plant is cut off from its water supply. How does this happen?

The water-absorbing cells of roots are called *root hairs*. Root hairs are cells at the tips of roots that bear the full responsibility for absorbing water and dissolved minerals from the soil. From root hairs, tubes called **xylem** transport water and dissolved minerals upward to the rest of the plant. Root hair cells have tiny hair-like protrusions that reach into the surrounding soil matrix. The soil matrix is a mixture of soil, water, and tiny pockets of air. Like all cells, root hair cells need oxygen, and they absorb it from nearby air bubbles in the soil matrix. However, in a flooded environment, excess water in the soil drives oxygen from the spaces between soil particles. Without oxygen, root hair cells die and can no longer draw in water to supply the rest of the plant.

Without live root hairs, the root essentially becomes sealed off, and the plant has no other way to obtain water from the soil matrix. Therefore, flooded plants die from lack of water.

Flooding also causes the soil to become **toxic** to plants. Normally, oxygen in the soil bonds with iron,

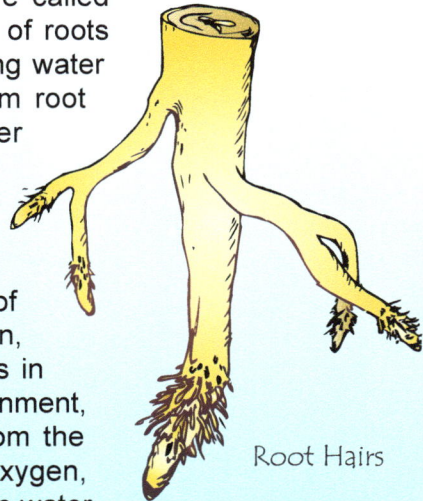

Root Hairs

Soil Particle

Air Bubble

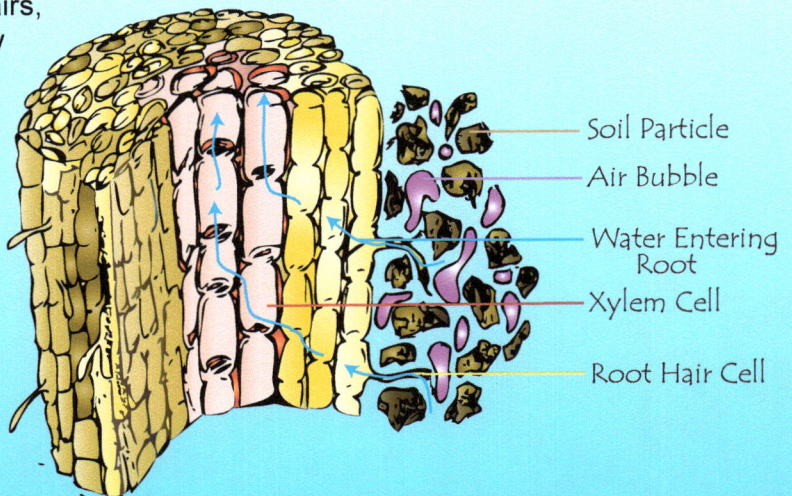

Water Entering Root

Xylem Cell

Root Hair Cell

Water Entering Root

magnesium, sulfur, nitrogen, and potassium. Plants can easily absorb and use minerals in this **oxidized** form. However, in flooded soils, microorganisms experience a limited oxygen supply and snatch the oxygen from oxidized compounds. Without oxygen, the minerals combine with other elements and become unusable or even toxic to plants.

However, plants that naturally live in wet places don't die! How are they able to survive in water when **upland plants** cannot? They have adapted strategies for overcoming the hazardous conditions of living in water. Plants described in the first part of this book live in or float on the surface of water. Formally, these water plants are called *aquatic plants*. The second part of this book describes plants that are rooted in the substrate with their upper stems and leaves out of the water. **Ecologists** call the latter ***emergent wetland plants***. The term *emergent* refers to the upper plant parts that grow above the water.

This book only considers **vascular plants**—that is, organisms in the plant kingdom with specialized **phloem** cells for conducting sugar and specialized xylem cells for transporting water and dissolved minerals. Plants without vascular systems transport materials from cell to cell without specialized conducting **tissues**. Non-vascular plants include **bryophytes** (mosses, liverworts, and hornworts). **Algae** and certain bacteria are other types of non-vascular **photosynthetic** organisms.

Adaptations for living in water evolved at different times and from unrelated groups of upland plants. To understand how plants adapted to living in water, we first need to understand the basic requirements of plants for gas exchange, exposure to light, structural support, and reproduction.

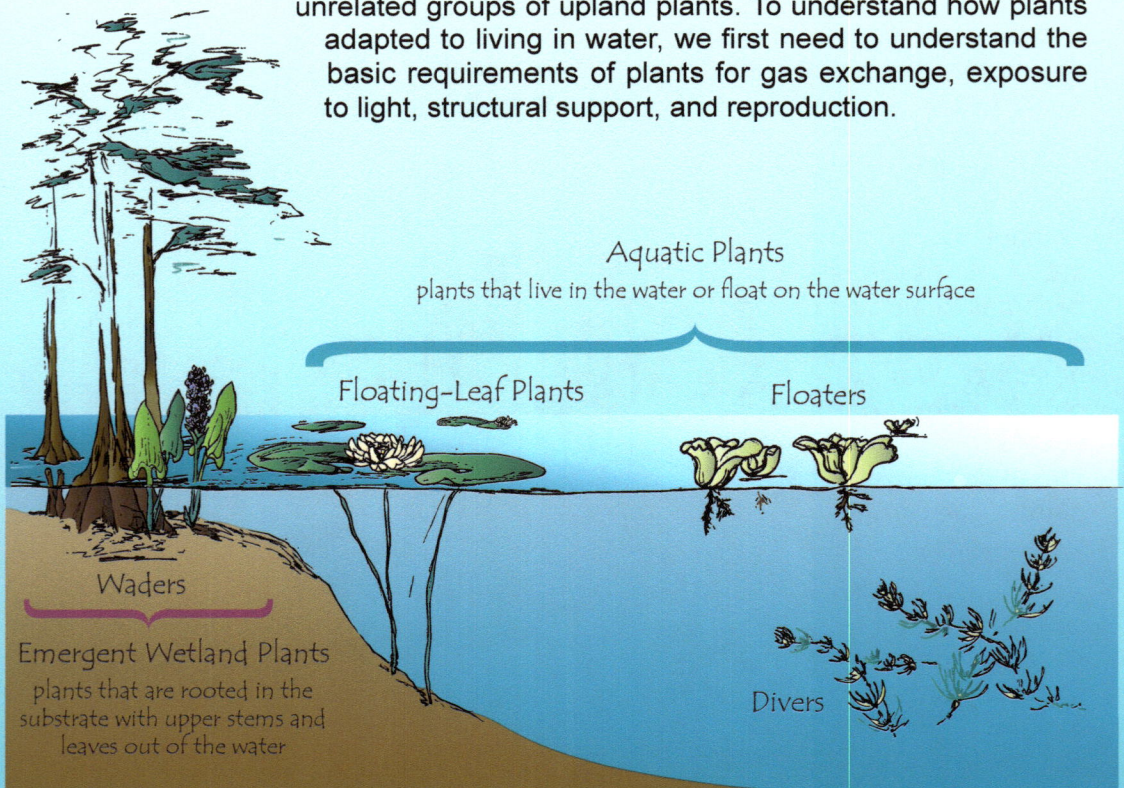

Aquatic Plants
plants that live in the water or float on the water surface

Floating-Leaf Plants

Floaters

Waders

Emergent Wetland Plants
plants that are rooted in the
substrate with upper stems and
leaves out of the water

Divers

Plants Need Gases

Plants need ways to exchange gases with their surrounding environment. Humans inhale O_2 and exhale CO_2 using lungs. Plants do the same but without lungs. Upland plants and some water plants have openings on leaf surfaces called **stomates** through which gases move in and out. Stomates consist of two **guard cells** surrounding an opening in the **epidermis**. When guard cells are **turgid** (tight) with water, they bend, causing the stomate to open and to allow gases to move in and out. During dry periods, guard cells shrink, causing the stomates to close. Closing the stomates reduces water loss, but it also inhibits gas exchange.

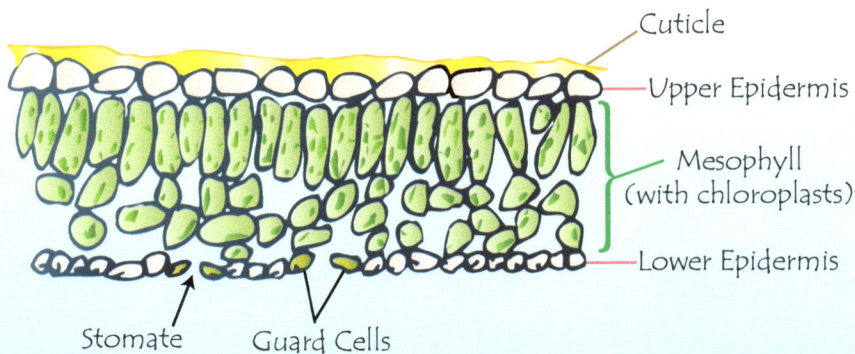

Leaf Cross-Section

In a single day, some plants can release up to several gallons of water into the atmosphere by way of their stomates through a process called **transpiration**, a type of evaporation from plants where plants shed water as a gaseous vapor. Although transpiration results in water loss, it serves the vital function of helping pull water from the roots to the tops of plants. To limit excess water loss, many plants (especially upland plants) have a waxy coating, called a **cuticle**, which covers cells that are exposed to air. The cuticle seals water inside cells.

In contrast to leaves, gases **diffuse** directly into and out of root cells without the use of stomates.

Oxygen gas (O_2) makes up about 20% of the air that surrounds us. Cells of most organisms use oxygen to break down food for energy, a process called *aerobic cellular respiration*. In **aerobic respiration**, the carbon (C) in food is combined with two atoms of oxygen (O_2) to form carbon dioxide gas (CO_2).

The important part of cellular respiration is converting food into energy. In this process cells use O_2, release CO_2, and store energy in chemical bonds of adenosine triphosphate, or **ATP**. *Triphosphate* means literally *three* *phosphates*. ATP, like money, is passed among cells to make things happen, to drive chemical processes. For example, ATP supplies energy to muscle cells when we move our arms. In spending energy on movement, each ATP used by muscle cells loses one of its phosphates. Energy is released when the phosphate's bond breaks. In our muscle example, this release of energy allows our muscle cells to contract.

Unlike animals, plants need carbon dioxide (CO_2). Although plant cells release CO_2 in aerobic respiration, they also use CO_2 in **photosynthesis**. In this process, carbon atoms from six CO_2 molecules link together to form the backbone of a simple sugar: **glucose**. Hydrogen and oxygen atoms bond to each carbon atom of this backbone to complete the glucose molecule. During photosynthesis, oxygen (O_2) is also released into the environment.

Most organisms depend on the oxygen produced by photosynthesis for aerobic respiration. A different type of respiration, **anaerobic respiration**, occurs without oxygen in some organisms and in some parts of the human body. Flooded plants can use anaerobic respiration for a short time. However, during anaerobic respiration, ethanol, a toxic alcohol by-product, accumulates in plant cells. In addition, aerobic respiration is far more efficient than anaerobic respiration, yielding 36 ATPs from each glucose molecule. Plants

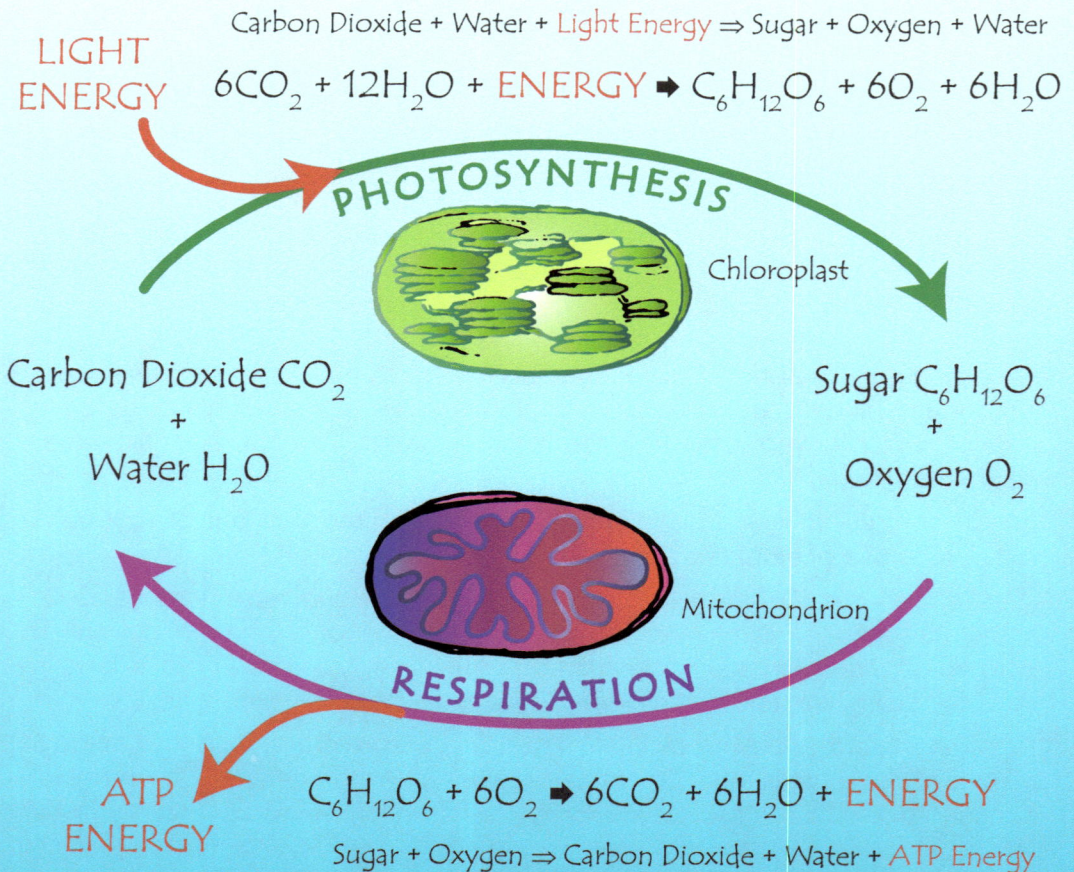

Carbon Dioxide + Water + Light Energy $\Rightarrow$ Sugar + Oxygen + Water

LIGHT ENERGY

$$6CO_2 + 12H_2O + ENERGY \rightarrow C_6H_{12}O_6 + 6O_2 + 6H_2O$$

PHOTOSYNTHESIS

Chloroplast

Carbon Dioxide CO_2
+
Water H_2O

Sugar $C_6H_{12}O_6$
+
Oxygen O_2

Mitochondrion

RESPIRATION

ATP ENERGY

$$C_6H_{12}O_6 + 6O_2 \rightarrow 6CO_2 + 6H_2O + ENERGY$$

Sugar + Oxygen $\Rightarrow$ Carbon Dioxide + Water + ATP Energy could not survive very long on anaerobic respiration alone, which produces only two ATPs from each glucose molecule.

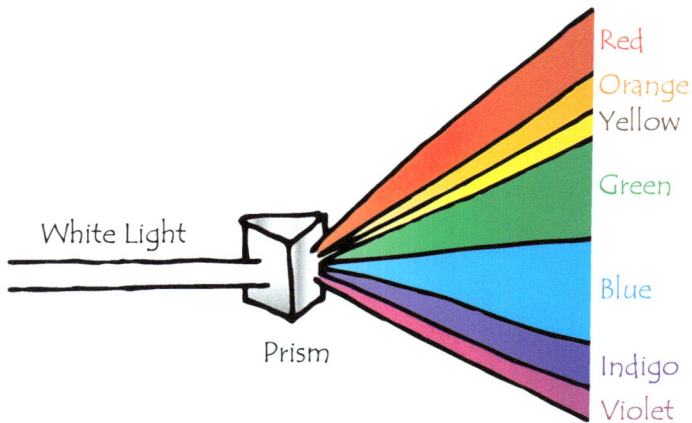

White Light

Prism

Red
Orange
Yellow

Green

Blue

Indigo
Violet

Plants Need Light

Plants, algae, and some species of bacteria need light to photosynthesize—that is, to convert carbon dioxide and water into glucose (sugar) and oxygen. In plants and algae, photosynthesis occurs in the chloroplasts, which are small green **organelles** (small organs inside cells). The green chlorophyll pigment in chloroplasts, absorbs sunlight. Yellow and orange pigments also absorb sunlight and transfer the energy to the primary photosynthetic pigment, chlorophyll.

To understand photosynthesis, we need to understand sunlight. When sunlight passes through a **prism**, it splits into a rainbow of colors. The rainbow in the sky is an example of the splitting of sunlight into colors—here, by rainwater molecules, which act as a prism. The array of colors visible to our eyes (red, orange, yellow, green, blue, indigo, and violet) accounts for a small part of all the radiation from the sun. A **mnemonic**, *Roy G Biv* (a boy's strange name), represents the colors of the visible spectrum, arranged from longest to shortest radiation **wavelengths**.

The wavelengths of light that are reflected off pigments give them the colors that we see. For example, the yellow we see in a painting is the result of paint pigments that reflect yellow. The chlorophyll pigment in plants, photosynthetic bacteria, and algae reflects green wavelengths (and absorbs red and blue wavelengths), so these organisms appear green.

What happens when chlorophyll absorbs light? Red and blue wavelengths of light cause the chlorophyll molecule to *lose an **electron***, which moves through the chloroplast as electrical energy. In other words, light from the sun is converted into electrical energy. This part of photosynthesis is similar to converting solar energy (sunlight) into electricity for household and industrial purposes using solar panels. Plants create electrical energy much more easily than do our solar power systems. Scientists study photosynthesis for clues about how to improve our use of sunlight to make electricity.

Electrical energy is used to make ATP molecules and another energized chemical compound called **NADPH**. In other words, *light* energy is converted into *electrical* energy, which is then converted into *chemical* energy in the form of NADPH and ATP. Because the manufacture of these compounds depends on light, we call this process the *light reaction* of photosynthesis.

Once ATP and NADPH are created, light is no longer needed, but photosynthesis is not finished. The final step creates a glucose molecule. This process is called the *dark reaction* of photosynthesis because it does not require sunlight.

Wait! What happened to the chlorophyll molecule that lost an electron in the light reaction? The chlorophyll grabs a replacement electron from one of the hydrogen atoms in a water molecule (H_2O). When chlorophyll takes an electron from hydrogen, the water molecule falls apart, and oxygen gas is released into the atmosphere.

In summary, photosynthesis converts twelve water molecules and six carbon dioxide molecules (in the presence of sunlight and chlorophyll) into a six-carbon glucose, six oxygen molecules, and six water molecules. Or, written as a balanced chemical formula:

$$12H_2O + 6CO_2 \rightarrow C_6H_{12}O_6 + 6O_2 + 6H_2O$$

Plants convert glucose into other **carbohydrates** such as fructose, sucrose, starch, and **cellulose**. We use sucrose (table sugar) to sweeten food and drinks. Table sugar is manufactured by processing sugarcane stems and sugar beet storage roots. Starch consists of several hundred glucose molecules linked together. Starch is the main ingredient in potatoes, rice, wheat, cassava, and other vegetables.

We call photosynthetic organisms (including plants, algae, and many groups of bacteria) *producers* because they use energy from a primary source, sunlight, to make carbohydrates. *Consumers* include all other organisms that feed on plants, plant products, or on each other. Without producers, consumers could not survive.

Plants Need Support

Humans stand upright because we have a skeleton of bones and attached muscles. Insects, crabs, and shrimp have an external skeleton, called an *exoskeleton*, that supports them. The most basic support for plants is the wall of cellulose surrounding each cell.

Cellulose is a network of strong fibers made of linked glucose molecules. Molecules, such as **lignin**, add strength by filling in gaps between the fibers of cellulose in cell walls. The cellulose fibers from cultivated cotton plants are used to manufacture thread for making cloth. The lignin in xylem is responsible for the hardness and strength of wood.

Many wetland and water plants also have spacious, air-filled layers of **aerenchyma** in their leaves, stems, or roots. Aerenchyma provides support and stores gases.

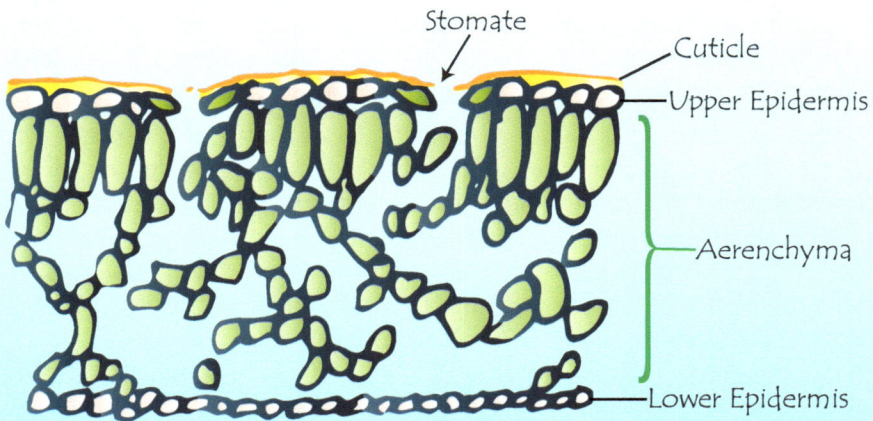

Stomate

Cuticle

Upper Epidermis

Aerenchyma

Lower Epidermis

Water Plant Leaf Cross-Section

Plants Need to Reproduce

Plants can be divided into seedless and seed bearing groups. Seedless plants include mosses, liverworts, ferns, and their relatives, such as horsetails. Seed plants are either flowering or non-flowering. Flowering plants use flowers for sexual reproduction. Most non-flowering seed plants are **cone** bearing trees and include pine, cypress, juniper, hemlock, spruce, cycads, and many others.

Ferns reproduce asexually by **spores** that **germinate** into separate, much smaller plants, which go unnoticed unless you know how to find them. During this phase of its life cycle, the fern reproduces sexually by making sperm and eggs that unite to form **embryos**. The sperm of most ferns swim through water to reach the egg. Embryos form after fertilization and grow into a new generation of larger plants.

Seeds are especially suited to drier conditions. They have a seed coat for protection, **endosperm** for food, and an embryo to grow into a new plant. The endosperm is a compact tissue of cells chock full of starch to feed the growing embryo as the seed germinates. By the time the endosperm is used up, the new plant has formed chlorophyll and can make its own food.

Pollen is produced by seed bearing plants and is the male part of the life cycle. Pollen germinates on the female part of the plant and grows a long *pollen tube* that contains sperm to fertilize eggs. The pollen tube is an evolutionary advancement for drier environmental conditions; it eliminates the necessity for swimming sperm. Still, wind, water, or animals must transport pollen grains to the female reproductive part—the **stigma** in flowering seed plants and the cone in non-flowering seed plants.

Parts of a Flower

Most non-flowering seed plants have pollen and cones designed for wind-**pollination**. The pollen grains have pockets for catching the wind, and the shape of the cone controls air currents around it. Pollen grains sail by, are caught in these air currents, and are pulled into the cone.

In flowering seed plants, the **anthers** (part of the **stamen**) produce pollen, and the ovaries (part of the **pistil**) bear eggs. Pollen grains land on the stigma and travel down the **style** to reach the eggs. If flowers contain stamens but no pistil, we call them **staminate**. If flowers contain a pistil but no stamens, we call them **pistillate**.

Animal **pollinators** such as insects, birds, and bats, are often attracted to flowers by their color or odor. Pollinators harvest food rewards of **nectar** (which contains sugar) and/or pollen (which contains proteins) from the flowers. Pollinators become dusted with pollen grains, and in visiting other flowers (or the same flower), pollen brushes off onto stigmas.

Wind-pollinated flowers lack bright colors, and the floral parts are reduced to prevent them from interfering with pollen movement. Anthers are enlarged to produce more pollen than in insect-pollinated flowers, and the pollen is often small, smooth, and light—features that facilitate wind transport. The timing for release of pollen in wind-pollinated plants coincides with dry weather periods. In wind-pollinated flowers, male and female parts are often in separate flowers. For example, corn plants have male tassels at the tops of stalks. The female ears of corn with long silks are lower on the stalk. Corn pollen grains land on the silks, which are the styles of the young corncob. Pollen grains germinate and grow tubes through the tissue of the silks and into the egg sacs. Each pollen tube releases sperm. The kernels on a cob of corn are seeds, comprised of an embryo, sweet or starchy tasting endosperm, and a seed coat.

Pollination occurs on the water surface or underwater in only 125-150 species of flowering plants (approximately 0.06% of all flowering plant species and 2-3% of all water plant species). The pollen of these species is not harmed by water. In the remaining flowering plants, if pollen gets wet, it can burst or grow an unsuccessful premature pollen tube. In addition, wet stigmas may rot or become incapable of receiving pollen. To protect flowers from getting wet, flowers often bloom during dry seasons, hang facing downward, or close at the approach of rain. Coverings of hairs or scales can prevent water from entering flowers.

Asexual reproduction does not require pollination or fertilization. Instead, part of a plant separates and grows into a clone of the parent plant. Some plants are also able to reproduce asexually by seeds formed without pollination and fertilization in a process called apomixis. Not all plants are able to clone themselves; however, **cloning** is common among water plants.

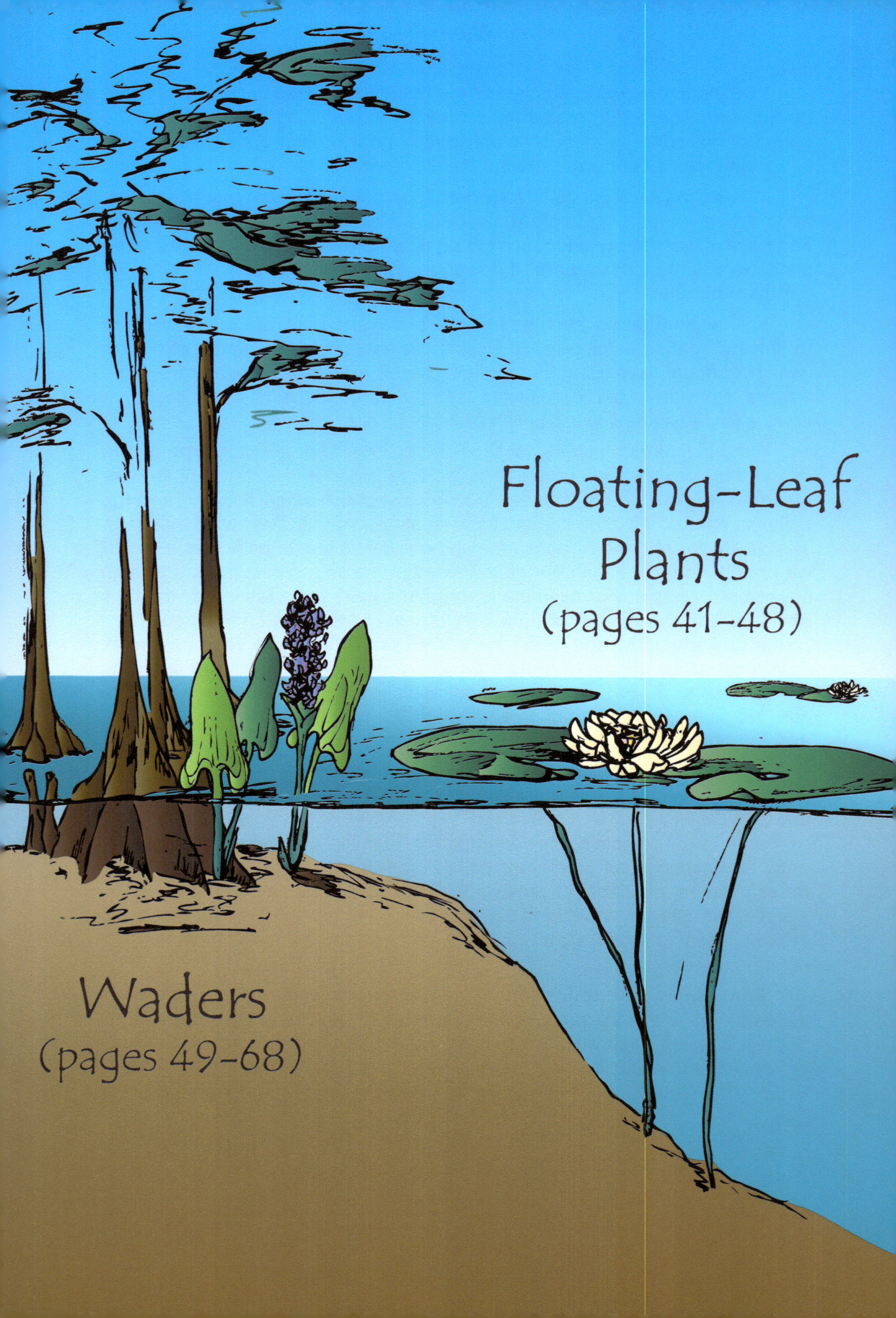

Floating-Leaf
Plants
(pages 41–48)

Waders
(pages 49–68)

Floaters
(pages 31-39)

Divers
(pages 15-30)

DIVERS

(SUBMERGED WATER PLANTS)

Divers live with all their body parts submerged underwater, except for the flowers of some species. Because the outer cells of Divers are surrounded by water, Divers have no need to protect themselves against water loss. Over time, waxy cuticles common to their upland plant ancestors became reduced or were eliminated. Divers are no longer able to survive on land; their leaves quickly dry and shrivel outside of water.

Because Divers are completely submerged, they face problems that Floaters, Floating-Leaf Plants, Waders, and upland plants do not. Specifically, there is less light, oxygen, and carbon dioxide in the water than in the atmosphere.

Many species of Divers are easily grown in aquaria with proper lighting and water clarity. Divers provide oxygen for aquarium fish, and the fish provide carbon dioxide for the plants. Fish also provide fertilizer for rooted Divers.

Divers: Gases

Divers absorb oxygen (for aerobic respiration) and carbon dioxide (for photosynthesis) directly from the water. In addition, rooted Divers absorb gases from the soil. Because the leaves of Divers are very thin, most of their cells are directly exposed to the water, enabling them to absorb gases without stomates.

Gases diffuse about 10,000 times more slowly into water than into air. Carbon dioxide, which represents only about 3% of the atmosphere, is in particularly short supply in the water, especially during the day when CO_2 is rapidly extracted from the water by photosynthetic organisms.

Divers and many other water plants store gases in reservoirs of aerenchyma tissue, groups of cells with large air spaces between them. At night, carbon dioxide produced in cellular respiration accumulates in these spaces. Aerenchyma can contain 100-500 times more carbon dioxide than the surrounding water and atmosphere. Carbon dioxide is also present in the soil, where it diffuses into roots and is transported to photosynthesis sites. During the day, oxygen from photosynthesis is trapped in aerenchyma and diffuses into roots and **rhizomes** where oxygen is limited.

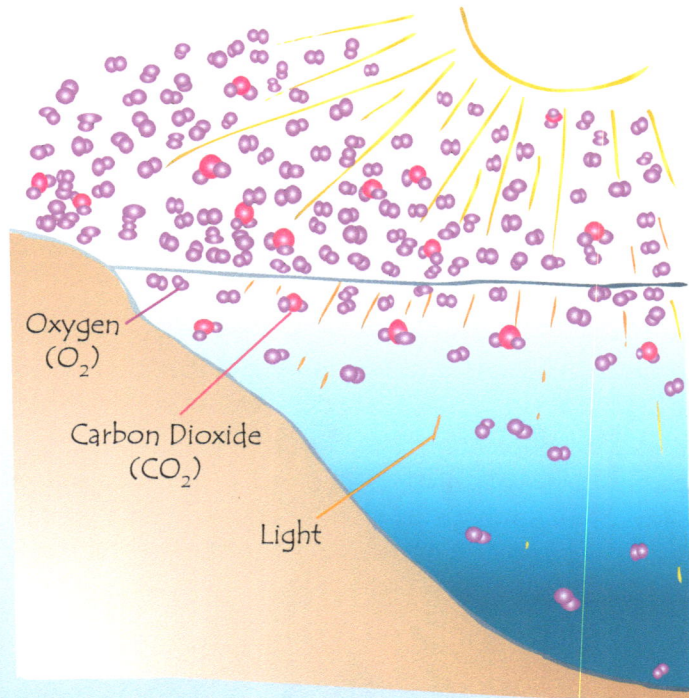

Light and Gases Decrease with Water Depth

Oxygen
(O_2)

Carbon Dioxide
(CO_2)

Light

Divers: Light

With limited light underwater, Divers must grow quickly toward the surface where light is more intense. When other plants dominate the surface, Divers grow toward openings to avoid being shaded. Some very successful Divers, such as hydrilla, Eurasian water milfoil, and pondweed, grow in thick canopies near the surface, shading other Divers.

The thin, ribbon-like or divided leaves of Divers allow more light to penetrate photosynthetic cells than other types of leaves. Unlike upland plant leaves, whose photosynthetic cells are deep within the leaf, the photosynthetic cells in Diver leaves are concentrated in the upper layer of the epidermis. Also, the total amount of chlorophyll in chloroplasts is higher in Divers than in other plants. This increases the rate of photosynthesis when light penetration is low.

Divers thrive only at depths of 12-17 meters (39-56 feet) below the surface in clear water. In murky, dark water, they must grow closer to the surface.

Divers: Support

Divers are supported by water; out of the water, they are limp and flaccid. They lack lignin and other strengthening materials that allow upland plants to grow stiff and upright. The narrow blades of fanwort leaves allow water currents to flow through without tearing them. The ribbon-like leaves of wild celery trail freely in the water, providing no resistance to currents.

One of the reasons Divers are popular for growing in aquaria is their buoyancy. Trapped air in the soft, spongy layers of aerenchyma in the roots, stems, and leaves, allows Divers to "float" in the water column instead of sinking into the darkness of deep water.

Divers: Reproduction

In the majority of Divers, as in other water plants, sexual reproduction requires flowers to be out of the water for insect pollination. Divers have various buoyant structures that hold their flowers above the water surface to keep them dry and attract pollinators. For example, bladderworts have aerenchymous arms at the base of their flowering stalks, which elevate the flowers above the water. One to three small, undivided, floating leaves hold fanwort flowers above the water.

Coontail, unlike most flowering species, *requires* water to carry its pollen from one flower to another. Huge amounts of pollen are released underwater, and both pollen and stigma are wet during pollination. In wild celery, pollen glides over the water surface to meet the stigma. Although water-pollinated flowers can often **self-pollinate**, most water-pollinated plants have staminate and pistillate flowers on different plants, thus ensuring **cross-pollination**.

Some Divers reproduce primarily through asexual **propagules** that separate from the parent plant and grow into new plants. Propagules also anchor the new plant to the substrate and store carbohydrates. Coontail reproduces mostly through asexual cloning. Birds, mammals, and even humans and their boats inadvertently carry asexual propagules, often dispersing plants to new locations.

Examples of Divers

Curly Leaf Pondweed (*Potamogeton crispus*), Clasping Leaf Pondweed (*Potamogeton perfoliatus*)

In several of the 90 species of pondweed, flowers are pollinated by "bubble self-fertilization." Air spaces in the flowers fill with gases that build up in the anthers. Pollen is released and transported on the surface of the bubbles. Each bubble increases in size until it touches a stigma. Pollen is then released from the bubble surface onto the stigma.

Most species of pondweeds have narrow, ribbon-like, or lance-shaped underwater leaves, and some have broad floating leaves. Most pondweeds are strict Divers with only underwater leaves.

Curly leaf pondweed, introduced from Eurasia, is a popular aquarium plant and has become a troublesome invasive species throughout North America. Native pondweeds provide important food and shelter for wildlife in **freshwater** habitats. Species are found throughout the United States, Canada, Mexico, and the West Indies.

Curly Leaf Pondweed
(*Potamogeton crispus*)

Water Milfoil (*Myriophyllum* species)

Water milfoil leaves are divided into narrow feathery segments. The flowers are barely noticeable to humans and are held above the water at the base of leaves. Water milfoils produce cyanide-containing compounds that deter grazers. Water milfoil species can be found throughout the United States and parts of Canada.

The Eurasian milfoil (*Myriophyllum spicatum*) and parrot feather (*Myriophyllum aquaticum*) are two of the most noxious **invasive** species in the United States. These plants have high photosynthetic efficiency and rapidly absorb nutrients from sediments and water. By repeated fragmentation, a single plant of Eurasian milfoil is theoretically capable of producing 250 million clones over a lifetime.

Fanwort (*Cabomba caroliniana*)

Fanwort gets its name from the shape of its underwater leaves. Each leaf is split into narrow fingers, like the open blades of hand-held folding fans. Fanwort plants have slender roots that anchor them in the soil, but the plants can also survive floating freely in the water. Fanworts have small flowers with white petals and pale yellow centers that attract insect pollinators.

The attractive leaves make this and related species popular for growing in aquaria. Fanworts grow in quiet waters from Massachusetts to southern Florida; westward to Tennessee, Kentucky, southern Illinois, Missouri, southeastern Oklahoma, Texas, and the western United States; and north into Canada.

Fanwort (*Cabomba caroliniana*)

21

Bladderwort, *Mousse d'Ecrivisse*—Crawfish Moss, *Etoile de Négresse*—Black Woman Star (*Utricularia* species)

Bladderworts are among approximately 500 species of **carnivorous plants**—those that trap and feed on other organisms. Thirty-five percent of all carnivorous plants occur in Australia, including bladderworts. Most carnivorous plants grow in wet places but are not necessarily water plants.

Bladderworts have bladder traps (or sacks) that capture small swimming animals. Bladderworts occur over a wide geographic range but are most common in areas where nitrogen is limited. Because Diver bladderworts are rootless, they must absorb nutrients from the water column. The animals that bladderworts consume provide nitrogen in the form of protein.

Bladder traps are positioned along the stems and among the leaves. At the entrance of the trap, a flap opens inward. Next to the trap are sensitive hairs that are stimulated by movement in the water, causing the trap to open, and drawing animals into the bladder. The flap closes the prey inside, and glands in the bladder pump the water out. Bacteria and digestive enzymes inside the bladder decompose the prey. At times, the bladders may become black, which is probably the reason for the name *black woman star*.

In a few species, floating branches hold flower stalks out of the water or wet mud. Like the spokes of a bicycle wheel, the floating branches join at the center where the flower stalk emerges.

Bladderwort (*Utricularia sp.*) "Swallowing" *Daphnia*

22

Coontail (*Ceratophyllum demersum*)

Coontail is a Diver that is common throughout North America, Tropical America, and Europe. It often grows with other Divers such as fanwort and bladderwort. The "coon" of coontail probably refers to the stems, bushy with leaves, that look like a raccoon's tail. The divided leaf segments are stiff and narrow with tiny pointed spines, so the plants feel rough to the touch.

The rootless nature and bushy growth of coontail make it a favorite for aquaria. It is easy to grow and multiplies by breaking into pieces to form new plants. Coontail also provides places for young aquarium fish to hide.

Coontail is an ancient flowering plant and probably evolved before most water plant species. The primitive flowers have no sepals or petals. Flowers rarely produce seeds because they are seldom pollinated. Coontail is water-pollinated, but because water transports pollen ineffectively, most of the pollen is usually lost.

Coontail (*Ceratophyllum demersum*)

23

Widgeon Grass (*Ruppia maritima*)

Widgeon grass provides an important habitat for animals in brackish water and **saltwater** worldwide. The leaves are thread-like, and the stems may be very long. Pollen grains are transported on the water surface in chains that float inside a mat of slime. The slime protects the pollen from getting wet. By forming chains, pollen has a better chance of reaching stigmas. In slow-moving streams, pistillate flowers of widgeon grass move back and forth with the current, collecting floating pollen. This species, like the pondweeds, may also be bubble self-pollinated.

Sea Grasses: Manatee Grass (*Cymodocea filiformis*), Shoal Grass (*Halodule beaudettei*), Tape-Grasses (*Halophila* species), Turtlegrass (*Thalassia testudinum*)

Sea grasses form important nursery grounds, providing shelter and food for fish, shrimp, lobsters, and other animals living in warm, shallow, marine waters. Sea grasses are water-pollinated, and they account for over half of all water-pollinated species.

Manatee grass is named for the manatee, a large vegetarian marine mammal. The plants have long, slender, straw-shaped leaves with pointed tips. Manatee grass grows along the coasts of the Atlantic Ocean, the Gulf of Mexico, and the Caribbean Sea.

Shoal grass depends on tides to carry pollen to stigmas. Long filaments of anthers are released in cottony masses on the water surface, forming large rafts of pollen. Exposed stigmas of pistillate flowers float, and when the tide rises, they collide with the rafts of pollen. Shoal grass grows in the Atlantic Ocean from North Carolina to Florida; in the Gulf of Mexico from Florida to Texas; and in the shallows of the Pacific Ocean.

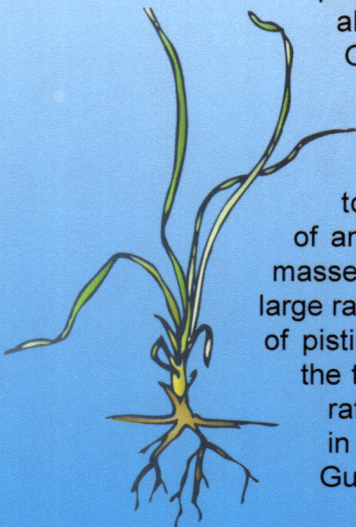

Manatee Grass
(*Cymodocea filiformis*)

Shoal Grass (*Halodule beaudettei*)

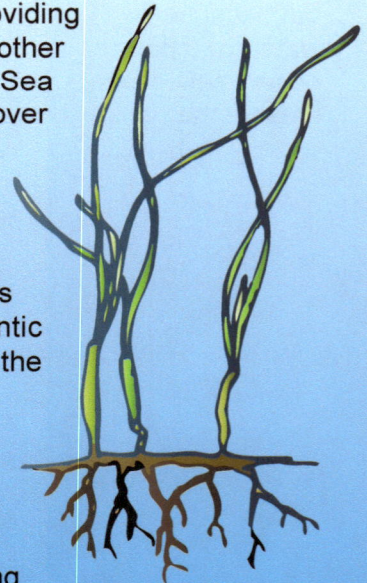

The Latin meaning of *Halophila,* the **genus** of tape-grasses, is *salt loving.* The genus is one of the largest of sea grasses worldwide and includes fourteen species found in tropical, subtropical, and temperate waters. In North America, *Halophila* plants grow in shallow waters of the Gulf of Mexico from Florida to Texas, the Bahamas, and the West Indies. Star grass (*Halophila engelmannii*) is named for the arrangement of leaves at the tips of branches. Paddle grass (*Halophila decipiens*) is the most widespread *Halophila* species, occurring in most oceans and seas worldwide. In Hawaii, paddle grass is an invasive species.

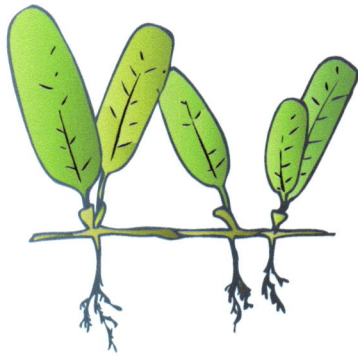

Paddle Grass (*Halophila decipiens*)

Johnson's seagrass (*Halophila johnsonii*), the only marine plant listed as threatened, occurs along the Atlantic coast of Florida from Sebastian Inlet near Melbourne to Virginia Key near Miami.

Turtlegrass has long, slender ribbon-like leaves that grow from thick underground stems. Flowers are produced a few centimeters from the soil surface during spring tides. Pollen forms strings of mucilaginous slime that sinks to the stiff stigma bristles of pistillate flowers. Meadows of turtlegrass create important habitats for animals from Florida to Texas and in South America, Bermuda, and the West Indies.

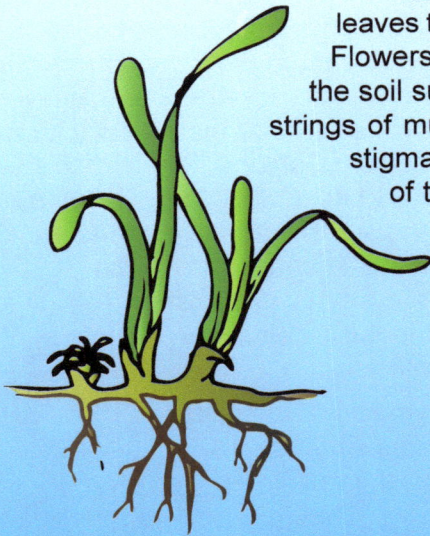

Turtlegrass (*Thalassia testudinum*)

Water Nymph (*Najas species*)

Water nymph pollen grains germinate before they are released into the water. Pollen grains and growing pollen tubes drift onto stigmas. Water nymphs produce abundant seeds that are important food for waterfowl.

Water nymphs grow in fresh and brackish waters of the Atlantic and Gulf states and in California and Oregon. Water nymphs also grow in Canada, Mexico, Central America, and the West Indies.

Hydrilla (*Hydrilla verticillata*)

Hydrilla has invaded all continents except Antarctica and is one of the most noxious invasive species in the world.

Two types of *Hydrilla* were introduced into North America, where they are thought to only reproduce asexually in nature. The first type, introduced around 1955, has only female flowers and has spread throughout the southern United States. The second, with both male and female flowers, was discovered in the late 1970s in the Potomac River. It appears to be more cold tolerant and is the most common type found north of South Carolina. Plants of this second type could theoretically reproduce sexually because (1) they have both male and female flowers; (2) experimentally raised plants have produced seeds; and (3) seed production is highly successful in other parts of the world. However, scientists have not found their extremely small seeds in natural settings in the United States thus far.

Plants spread through asexual propagules called **turions** and **tubers**. Turions are formed at the bases of leaves and the tips of stems. They fall to the substrate, overwinter, and sprout in the spring. Tubers form underground. Millions of turions and tubers are found in the sediments of *Hydrilla*-infested waters. In a Florida canal, researchers counted an average of 918 turions per square meter. By the mid-1960s, *Hydrilla* had become the most abundant Diver in Florida.

Hydrilla Turion

Hydrilla Tuber Sprouting

In Southeast Asia, where *Hydrilla* is native, it reproduces sexually. Staminate flowers are released underwater and float to the surface. Pollen shoots explosively into the air. A small percentage of pollen falls on stigmas of pistillate flowers. Staminate flowers may also sail over the water and bump into pistillate flowers. If rising water covers pistillate flowers, the petals and sepals close around an air bubble to prevent the stigmas from getting wet.

Hydrilla (*Hydrilla verticillata*)

South American Elodea (*Egeria densa*), Canadian Pondweed (*Elodea canadensis*)

South American elodea originated in South America. However, aquarium enthusiasts, fond of its attractive leaves and hardy growth, have distributed the species worldwide. Plants grow so thickly that waterways become impassable to motorboats. It is considered one of the most troublesome invasive species. In many places, plants cannot be purchased to grow in aquaria for fear of further spread into natural bodies of water.

Leaves and whole plants of South American elodea and Canadian pondweed are used in classes to demonstrate the production of carbon dioxide in photosynthesis. Also, because the leaves are thin enough for light to pass through them, they are used to study cell structures. When looking through a microscope, one can see green chloroplasts moving around large **central vacuoles** inside the cells.

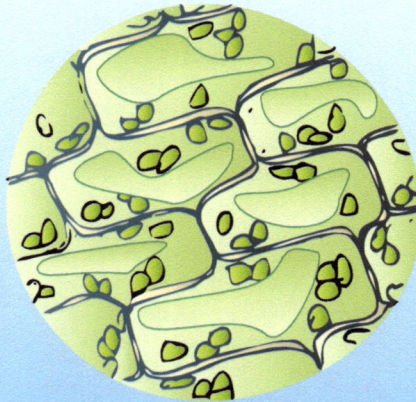

Elodea Leaf Cells

Pistillate flowers of South American elodea grow at the ends of long stalks. Staminate flowers are released as buds and spray pollen over the water surface. Pistillate flowers form a depression in the water surface, which draws pollen grains to the stigmas.

South American Elodea (*Egeria densa*)

Canadian pondweed (*Elodea canadensis*), native to North America, is often confused with South American elodea (*Egeria densa*) and hydrilla (*Hydrilla verticillata*). Their similar names create even more confusion.

Egeria densa

Elodea canadensis

Hydrilla verticillata

29

Wild Celery (*Vallisneria americana*)

Wild celery has long ribbon-like leaves that grow from the base of the plant and trail through water currents. In soils that are deficient in phosphorus, a **symbiotic mycorrhizal fungus** associates with wild celery root tips. The fungus modifies phosphorus-containing chemical compounds in the soil to create more usable forms for the plant. The fungus, in turn, acquires carbohydrates from the plant.

In sexual reproduction, large numbers of male flowers (about 1 millimeter or 0.04 inches in diameter) are released from underwater sacks. The male flower petals surround a gas bubble that allows them to rise to the surface where they open and float. The large female flower creates a dimple on the water surface that the male flowers slide into. After pollination, the stalks of the female flowers coil spring-like and pull the flowers underwater where they develop many-seeded fruits.

Wild celery is one of the most popular plants for aquaria. Its natural habitat is freshwater throughout the eastern and southern United States, as well as New Mexico, Arizona, North and South Dakota, and Canada.

Water Crowfoot, Buttercup (*Ranunculus flabellaris* and other *Ranunculus* species)

Water crowfoot has leaves with many thread-like or narrow ribbon-like divisions as well as leaves that float on the surface. A cuticle protects the floating leaves, which allows the plants to survive low water conditions.

The attractive yellow or white *Ranunculus* flowers attract insect pollinators. Water crowfoot is particularly common in the northeastern and midwestern United States but can also be found in Texas, New Mexico, California, and Canada.

FLOATERS

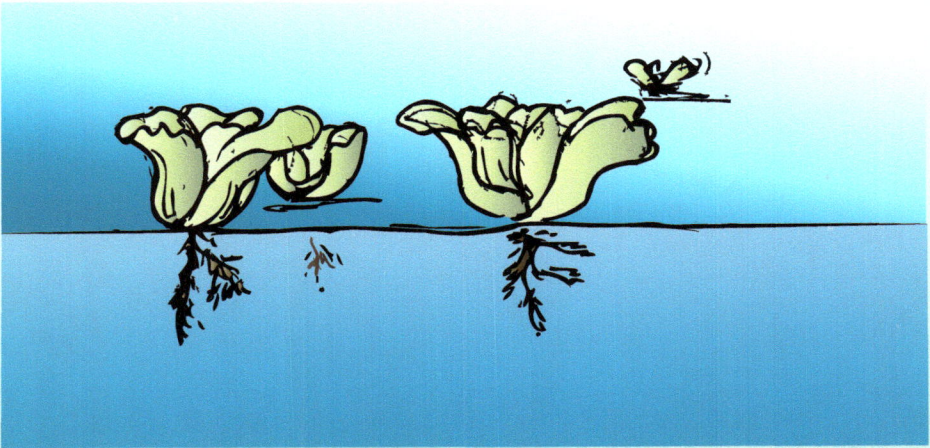

Floaters float on the surface of the water. Wind and water currents carry Floaters along the water surface. Roots hang freely in the water to absorb water and nutrients and to stabilize the plant. Occasionally, plants that drift ashore may take root in soil. In Floaters such as duckweeds, only the upper leaf surface is exposed to the atmosphere. Cells on the lower surface absorb water directly. Floaters reproduce extensively by asexual reproduction.

Floaters: Gases

Floaters have retained many features of upland plants. A cuticle covers exposed epidermal leaf cells, and stomates exchange gases. Roots release large amounts of oxygen into the water. The extensive aerenchyma tissue that traps and transports gases makes the plants feel soft and spongy.

Floaters: Light

As in upland plants, photosynthetic cells of Floaters are exposed to direct sunlight.

Floaters: Support

The extensive spongy aerenchyma of Floaters provides buoyancy. On the underside of American frog's bit leaves, a patch of spongy cells, like miniature bubble-wrap, keep the plants afloat. Floaters differ widely in the amount of their structural support. Duckweeds, the smallest Floaters, are supported entirely by water. However, in larger plants, internal structure is needed to hold them upright and stiff above the water surface.

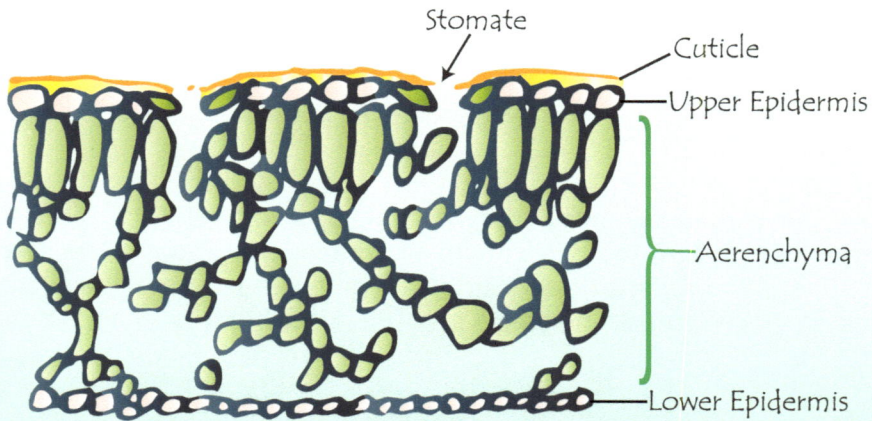

Water Plant Leaf Cross-Section

Floaters: Reproduction

Floaters reproduce both sexually and asexually. Flowers are held above the water to attract pollinators. In water hyacinth, the flowers are large and beautiful and produce nectar and pollen to reward insect pollinators. Water lettuce flowers release a strong rotten meat odor on sunny mornings to attract fly pollinators. Water lettuce flowers offer no food rewards but may provide resting places on the water for flies. The flies get pollen on their feet and carry it from flower to flower.

Duckweed flowers are the smallest among flowering plants. In *Wolffia*, the flower can best be seen when magnified. Pollen grains float in the water from one flower to another to pollinate them. After pollination, each flower makes one seed.

Examples of Floaters

Duckweeds (*Lemna, Spirodela,* and *Wolffia* species)

Duckweeds have one, two, or three green **fronds** that lie flat on the water. Fronds are a combination of leaves and stems. The top of each frond is exposed to the air, and the underside is in water.

A single plant can fit on the end of a child's finger. *Lemna* and *Spirodela* have roots that grow into the water from the center of each frond. *Lemna* has only one root per frond; *Spirodela* has two or more. *Wolffia* is a tiny green bead without roots. It is able to float due to the surface tension of the water.

Masses of small duckweeds form floating green carpets on waters of sluggish bayous, ditches, lakes, and **swamps**. Where water is rich in nutrients, they cover the surface, shading out plants beneath them. Several species of duckweeds often grow together. They rarely flower but reproduce asexually by forming new fronds that break away from the parent plant.

Duckweed on Child's Hand

Water birds, fish, crawfish, and nutria graze on duckweeds. Nutria, a vegetarian mammal, dips its furry chin into the water and scoops duckweeds into its mouth with its two front paws. These animals were introduced into Louisiana from South America for their fur in 1938. Since then, they have overpopulated wetland areas and become nuisances. Their very large populations can destroy the plant life of marshes. Nutria burrow into the shorelines and undermine levees, roads and other structures around waterways.

Single *Lemna* Plant

Nutria Eating Duckweed

Water Hyacinth (*Eichhornia crassipes*)

In 1884, at the International Cotton Exposition in New Orleans, Louisiana, visiting members of the Japanese government gave away souvenir water hyacinth plants from Venezuela. The large, exotic multiple blooms with purple petals and a yellow bull's eye nectar guide were prized treasures. People took the plants to garden and farm ponds in surrounding districts of New Orleans and other parts of Louisiana. The plants quickly multiplied, and people threw them into the nearest lakes, ponds, and rivers to get rid of them.

Water hyacinth reproduces sexually and asexually. The flowers are insect-pollinated. In asexual reproduction, new clusters of leaves, called **rosettes,** are formed on the ends of long stems called **stolons**. Rosettes can easily break off and form new mats of plants. Water hyacinth has become one of the most noxious invasive species in tropical and subtropical areas. In a few days to two weeks, water hyacinth mats can double in size.

Leaf blades of water hyacinths catch the wind like sails and move the plants over the water. Rafts of plants can trap unsuspecting boaters and entangle propellers. State and federal governments in the southeastern United States spend billions of dollars every year to control the spread of water hyacinth.

Water Hyacinth (*Eichhornia crassipes*)

Rafts of floating water hyacinth are so huge and thick that they hide the water beneath them. Photosynthetic organisms underneath receive limited light and can produce little oxygen in the water. However, water hyacinths transport large amounts of oxygen into the water through their roots.

Although boaters consider water hyacinths pests, they are friends of other plants and animals. For example, seeds of the water-spider orchid (*Habenaria repens*) and water primrose (*Ludwigia* species) germinate and grow on mats of water hyacinths. The feathery, purple-tinged roots of water hyacinths provide habitats for many small animals.

By absorbing excess **nitrates** and phosphates in the water, water hyacinths help clean water that is **polluted** by household waste and fertilizer run-off. Plants grown in wastewater multiply rapidly and grow large.

Water Hyacinth Clogging the Waterways

Mosquito Fern (*Azolla caroliniana*)

Mosquito ferns have special pouches that house colonies of single-celled **symbiotic cyanobacteria**. The cyanobacteria convert nitrogen gas into a form of nitrogen that mosquito ferns can use. The ferns float in very large colonies, turning the surface of the water from red to blue-green and back again, depending on the water acidity. Chinese farmers encourage mosquito ferns because they provide fertilizer for their rice fields. Before planting, the farmers drain the fields, causing the fern plants to die, decompose, and release nitrogen fertilizers into the soil.

Plants and animals need nitrogen for making proteins and DNA but are unable to use nitrogen gas from the atmosphere. Plants depend on nitrogen-fixing bacteria to convert nitrogen gas into ammonia (NH_3), which plants can use. We call the process of converting nitrogen gas into a usable form **nitrogen fixation**. Most plants in the bean family, as well as many other species of plants, provide a home and food in **root nodules** where bacteria live and fix nitrogen. Free-living, soil-dwelling, nitrogen-fixing bacteria also release ammonia that plants can use.

Water Spider Orchid (*Habenaria repens*)

Water spider orchids can be classified as Floaters or Waders. As Floaters, they grow on floating mats of vegetation, such as rafts of water hyacinth. Many small green flowers form along a naked stalk. Orchid seeds lack the endosperm that supplies food for germinating seeds and seedlings in other species of flowering plants. For orchid seeds to germinate and for young plants to grow from embryos, they associate symbiotically with a mycorrhizal fungus. The relationship is symbiotic because it benefits both the fungus and the orchid. The fungus converts nutrients in the surrounding environment into usable forms for the young plant, and the plant provides carbohydrates for the fungus. Water spider orchids can be found from North Carolina south throughout Florida and west to Texas as well as in the West Indies, eastern and southern Mexico, Central America, and northern South America.

Water Lettuce (*Pistia stratiotes*)

Water lettuce has gray-green, velvety water-resistant leaves clustered in rosettes. Their long, feathery roots dangle in the water beneath them. Like water hyacinths, they grow in large colonies and become nuisances that quickly cover the surface of the water. They reproduce asexually by producing new plants at the ends of long stolons that easily break away from the parent. Water lettuce grows in fresh water from Florida to Texas and throughout the tropics. Like water hyacinths, these plants are used to clean polluted water.

Water Lettuce (*Pistia stratiotes*)

FLOATING-LEAF PLANTS

Floating-Leaf Plant leaves float on the surface of the water. Leaves of Floating-Leaf Plants, as well as many other types of water and non-water plants, consist of an expanded flattened leaf *blade* that connects to the stem via a stalk, or **petiole**. In Floating-Leaf Plants, long flexible petioles connect leaf blades to rhizomes anchored in the soil. Rhizomes are modified stems that grow horizontally underground or on top of the ground and are often pocked with knobby leaf scars (places where leaves were previously attached but have subsequently fallen off). The flat, round, heart-shaped, or oval leaf blades of Floating-Leaf Plants are tough, leathery, and waterproof. In water lilies, the floating leaf blades are called lily pads. The long, flexible petioles allow leaf blades to spread over the water surface. The large floating leaves shade the water beneath, limiting the growth of Divers below.

Floating-Leaf Plants: Gases

Gas exchange in Floating-Leaf Plants occurs entirely via stomates located on the upper surface of the leaf blade. A cuticle prevents the exposed surface cells from drying out. Spatterdock, American lotus, big and little floating-heart, sacred lotus, water lilies, and many other Floating-Leaf Plant species force gases through tubes, similar to the way a swimmer draws in and blows out air through a snorkel.

In these snorkeling plants, air is drawn in through the stomates on young leaf blades and forced downward to underground stems and roots. Unwanted or unneeded gases, such as carbon dioxide, are moved through the internal tubing that leads back up to older leaves. In spatterdock, the hollow passageways make up 60% of the petioles and 40% of the underground stems and roots.

Bacteria living in the soil around the roots of snorkelers produce **methane**, which is poisonous to plants and many other organisms. Roots of snorkeling plants take in methane and rid the soil of it through their internal passageways. Methane passes out of snorkeler plants through older leaves. Snorkelers benefit the soil by oxygenating it.

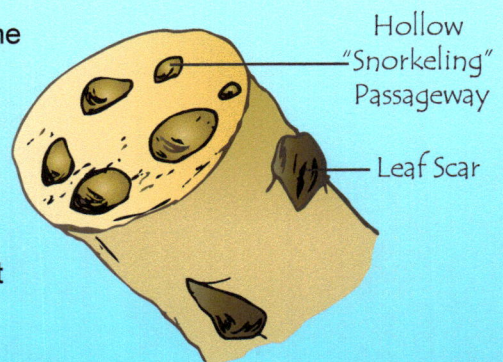

Hollow "Snorkeling" Passageway

Leaf Scar

Spatterdock Rhizome Cross-Section

Floating-Leaf Plants: Light

Interception of light by Floating-Leaf Plants is similar to that of upland plants. Light strikes the upper surfaces of the floating leaves where photosynthesis occurs. Because the floating petioles are flexible, leaf pads can move into open spaces where more light is available.

Floating-Leaf Plants: Support

The aerenchymous tissue in the petioles and veins of the floating leaves keep them afloat. Air spaces also soften the blows of rain on leaves. Rhizomes, lying buried or resting on the ground, are supported by the substrate.

Floating-Leaf Plants: Reproduction

Floating-Leaf Plant flowers are borne on long **peduncles** (flower stalks) and float on the surface, as in water lilies, or are held above the water, as in American and sacred lotus and spatterdock. The flowers are generally large, colorful, scented, and have ample pollen and nectar.

Asexual reproduction occurs via various types of propagules. The banana plant is named for its tubers that are shaped like bananas. These small tubers float in the water for a while, then sink and root in the soil. Multiplying by tubers is a form of cloning, and the new plants that grow from them are genetically identical to the parent plant.

Examples of Floating-Leaf Plants

Water Lily, *Herbe au Cœur*—Heart Plant, *Papayeur*—Paddler (*Nymphaea* species)

Water lilies are found in ponds and along the edges of lakes and bayous. Some species of water lilies are named for the colors of their flowers, which open wide like saucers at the ends of long stalks and float on the water surface. The flower colors, scents, nectar, and abundant pollen attract pollinators. After pollination, flowers sink, and each forms a pod of seeds. Water lilies produce poisonous **alkaloids** that deter grazers. Some are snorkelers.

Over 50 species of water lilies are distributed worldwide. Banana water lily is found in Mexico, Florida, Louisiana, Texas, and Arizona. White water lily is more widespread from Newfoundland and Quebec to Manitoba in Canada and southward into the United States to South Florida, Oklahoma, eastern Texas, and Arizona. Blue water lily grows in Mexico and in the United States in the peninsula of Florida, Louisiana, and southern Texas.

Water Lily (*Nymphaea sp.*)

Spatterdock, American Lotus, Yellow Lotus, Water Chinquapin, *Grains à Volée* (*Nuphar species*), Sacred Lotus (*Nelumbo nucifera*)

Spatterdock grows in open water along the edges of bayous, ponds, and lakes and produces alkaloids poisonous to grazers. Throughout its distribution in temperate areas of North America, Europe, and Asia, there are 10-12 species of spatterdock. Recent comparisons of spatterdock DNA samples suggest that there are probably eight species of *Nuphar* in North America.

Spatterdock leaves occur in three positions relative to the water surface: floating, submerged at the base of the plant, and (in some species) elevated out of the water. The large submerged leaves are thin like those of Divers. The flat, valentine-shaped, floating and raised leaf blades have shiny, waxed surfaces that shed water. Plant snorkeling (or *pressurized ventilation*) was first discovered in spatterdock leaves.

The single yellow spatterdock flowers reach above the water on peduncles that can be almost two meters (six feet) long. The center of each young flower contains a red-rimmed, greenish-yellow pod that resembles a bathroom shower head turned upward. Dozens of yellow, pollen-filled stamens surround the pod. The flowers are pollinated by several types of insects, after which, a many-seeded fruit develops above the surface of the water. As the seedpods mature and grow larger, they turn to the east and crook like a bishop's staff. By winter, the grape-sized seeds rattle inside the dark brown, woody pods and poke out of holes in the top.

Spatterdock (*Nuphar sp.*)

Grains à volée, which means "flying seeds," is the French name for spatterdock. Despite the name, the seeds are too heavy for wind to carry them. Perhaps the name refers to seeds flying in the faces of boaters who run into the ripe pods. The colonies of many thousands of plants amassed in lakes and quiet waters may be the remnants of abandoned Native American spatterdock farms. The nutty-flavored flesh inside the hard covered seeds is a nutritious food source.

The lemon-yellow flowers of spatterdock, and the pale pink flowers of sacred lotus are the largest and most spectacular of any water plant. Sacred lotus can heat its flowers to 30°C (86°F) for as long as four days, which can be as much as 20°C (36°F) higher than air temperature. Heating releases a scent from flowers that attracts insect pollinators. On the first morning, petals of the warmed flowers open slightly to expose the stigmas and tips of stamens. Pollinators may bring pollen from other flowers to cross-pollinate. That night the petals close, trapping the pollinators inside, and the stamens shed pollen into the bowl-shaped base of the flower to feed the pollinators. The next morning, the flower opens wider. Pollinators are freed, covered with pollen, to visit other flowers or to self-pollinate the same flower. The flower remains open that day, available for self- and cross-pollination of any unfertilized eggs. Seedpods develop above the water surface.

Native to India, sacred lotus is widely cultivated. Buried seeds of sacred lotus found in Manchuria were able to germinate after 200 years, and seeds may survive hundreds or even a thousand years.

Lotus (*Nelumbo sp.*)

Big Floating-Heart, Banana Plant (*Nymphoides aquatica*), Little Floating-Heart (*Nymphoides cordata*)

Floating-hearts have several attractive white to cream flowers with yellow centers that poke up near the notch of heart-shaped leaf blades. The undersides of the leaf blades of big floating-heart are purplish and rough. The name banana plant refers to the cluster of swollen tubers, which are asexual reproductive structures. Plants of big floating-heart or banana plant are favorites for aquaria, and "bananas" buried in the substrate can grow into new plants. The floating hearts are snorkelers. Little floating-heart is more widespread than big floating-heart, occurring from Florida to Texas, Canada, and New England. Big floating-heart grows from New Jersey to Florida and west to eastern Texas. There are approximately 50 species of *Nymphoides* distributed throughout the tropics.

Banana Plant (*Nymphoides aquatica*)

WADERS

(EMERGENT WETLAND PLANTS)

Waders (also known as *emergent wetland plants*) live with their "feet" in the water and the rest of their bodies above the water. *Emergent* refers to the plant parts that emerge above the water, leaving only the lower sections of their stems underwater. During dry periods or low tide in coastal areas, there may be little or no standing water around the base of Waders. To anchor the plants, roots burrow into the water-saturated soil under and around lakes, bayous, swamps, and rivers.

Some Waders are adapted to living in deeper water than others. Plants of some species root in wet soil and tolerate flooding for only short periods. Most are well adapted to flooding and fluctuating water levels. For example, when floods reduce gas availability in the soil, the upper parts of the plant transport gases to and from roots. Like Floating-Leaf Plants, many Waders are snorkelers, pumping gases from leaves to roots and back again.

Waders form aerenchyma for gas storage and subsequent transport to roots when necessary. In bald cypress, swamp maple, tupelo, gum, and other swamp trees, flooding triggers swelling or buttressing around the base of the trunk. **Buttresses** contain aerenchyma tissue, which not only stores and transports gases but also increases the stability of trees in water.

Many Waders form leaves of different shapes and sizes in response to fluctuating water levels. Underwater leaves of *Sagittaria* are sword-shaped, and aerial leaves vary among the species from narrow to arrow-shaped. In some species, rising water triggers plants to form floating leaves, which ensure survival under low water conditions when underwater type leaves would dehydrate.

Most wetland plant species are Waders. A characteristic example, bald cypress, lives in deep water swamps that may be flooded for 10-12 months of the year. Although these trees normally live in swamps, they also grow well when planted in uplands. For seeds to germinate, they must soak in water for extended periods. But the young seedlings will survive only if water recedes enough for leaves to be above the water surface.

Waders: Gases

A major challenge for Waders is getting oxygen to their roots. Leaves and green stems have stomates for gas exchange, but soil that is underwater is deprived of oxygen needed for survival of root cells. Some Waders have roots that can tolerate living under low oxygen conditions. Others store gases in aerenchyma and/or transport gases through snorkeler passageways. For example, common reed, cattails, and many other Waders are snorkelers. Much of the oxygen that is transported to the roots, diffuses out of the roots, creating a more oxygenated soil. Carbon dioxide moves upward from the roots to the leaves for use in photosynthesis.

Grasses and sedges living in saltwater and brackish waters are especially vulnerable to reduced gas exchange as a result of oil spills. When oil coats the leaves of marsh plants, their stomates become clogged and are unable to obtain or release gases. The concentration of gases dissolved in water significantly decreases when the surface is covered with oil. In addition, roots damaged from oil spills become less effective in transporting nutrients upward. During cleanup efforts, trampling of the marsh through walking and driving also damages the plants.

Waders: Light

Leaves of Waders must be above the water surface to absorb light and take in gases for photosynthesis. Rising water triggers rapid growth of shoots, and in some species such growth occurs within 30 minutes of flooding. Lengthening of shoots is caused by the elongation of cells rather than the multiplication of cells.

Waders: Support

Like upland plants, Waders are strengthened by the lignin that impregnates xylem walls.

Waders: Reproduction

Waders are wind- or insect-pollinated, and many species reproduce asexually. Wind-pollinated Waders include grasses, sedges, reeds, cattails, willows, bald cypress, and wild rice. Insect-pollinated Waders have large, colorful, scented flowers. In addition, one tropical mangrove Wader is bat-pollinated.

Irises are an example of insect-pollinated Waders. They have spectacular yellow, blue, purple, or brick red flower petals with a yellow patch in the center. The droopy flowers group in three's, resembling a single large flower. Insects find the bright yellow nectar guide patches that lead them to the nectar and pollen. In rummaging in the flowers for food, insects become dusted with pollen and carry it to other flowers. After the flowers have faded, a dark brown, woody pod forms. Inside the pods are three rows of rock-hard seeds.

Examples of Waders

Large or American Cranberry (*Vaccinium macrocarpon*), Small Cranberry (*Vaccinium oxycoccos*)

Cranberries are farmed for their sour, bright red to purplish berries that are used to make sauces and juices. These bushy shrubs grow in bogs in eastern Canada and in the northeastern United States, southward into the mountains of North Carolina, eastern Tennessee, and Arkansas. When the fruits are ripe and ready to harvest, farmers flood the cranberry bogs to dislodge the berries. The berries float, and the farmers scoop and push them into collecting containers.

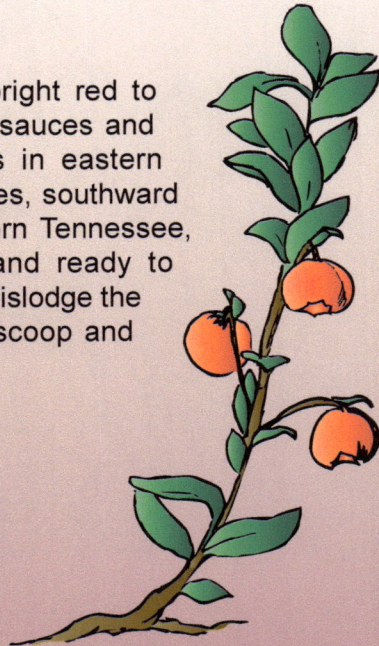

American Cranberry
(*Vaccinium macrocarpon*)

Papyrus (*Cyperus papyrus*)

Papyrus, a grass-like marsh plant in the sedge family, is native to Africa. Ancient Egyptians made the first paper from papyrus. The buoyant stems are also used for making boats. According to the biblical story, baby Moses was put into a floating basket among bulrushes. The bulrushes were papyrus plants growing along the margins of the stream. These attractive plants are widely cultivated as ornamentals in outdoor ponds.

Papyrus (*Cyperus papyrus*)

Sagittaria, Wapato, Bull Tongue (*Sagittaria lancifolia*), California Arrowhead (*Sagittaria montevidensis*), Grassy Arrowhead (*Sagittaria graminea*), Burhead (*Echinodorus* species)

The many species of sagittaria and burheads have flowers scattered on long, leafless stalks. Each flower has three petals, measuring 2.5-5 centimeters (1-2 inches) across. Most sagittarias and all burheads have white petals. Normally, the staminate flowers of the sagittarias have yellow centers, and pistillate flowers have green centers. An exception, California arrowhead, has a red spot at the base of its petals. The flowers of burheads are **perfect**— that is, they have male parts (stamens) and female parts (pistils) on the same flower. Yellow stamens at the base of the pistils ring the green centers of burhead flowers.

Several sagittarias bear small, white, potato-like tubers on their **runners** (horizontal stems that tunnel through the mud). Because ducks dig and eat the starchy "potatoes," we call these plants *duck potatoes*. The tubers taste like potatoes when roasted or boiled and were an important source of food for Native Americans. The tubers are asexual propagules that can survive over the winter and start new plants.

Bull tongue, named for its flat, wide leaf blade, grows in freshwater marshes and along the edges of lakes, streams, and swamps. Bull tongue can be found in the Atlantic and Gulf Coastal states.

California arrowhead occurs in temperate areas of North and South America but not in between. Migratory birds probably carried seeds between these isolated geographic areas.

There are over 25 burhead species, many of which are popular for growing in fish tanks. In nature, burheads have submerged and aerial leaves. However, if submerged in aquaria for long time periods, they grow only submerged leaves. Underwater leaves of some species are broad and oval, while others are long and thin. Aquarium enthusiasts refer to the species that have long, thin leaves as *sword plants*. Like burheads, sagittarias are frequently grown in aquaria, where they also produce sword-shaped leaves.

Over 30 species of sagittarias are distributed mostly throughout South, Central, and North America, with a few species in Europe and Asia.

Red Mangrove (*Rhizophora mangle*), Black Mangrove (*Avicennia germinans*), White Mangrove (*Laguncularia racemosa*), Buttonwood (*Conocarpus erectus*)

Mangroves are trees and shrubs found in tropical and subtropical coastal areas. They have adapted to survive in shallow saltwater where water levels fluctuate frequently. They eliminate the salt from seawater through leaf glands. Shiny salt crystals accumulate on the leaves and can easily be seen and tasted.

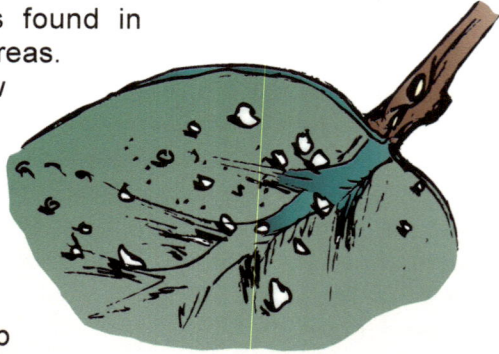

Mangrove Leaf Salt Crystals

Red mangroves have arching prop roots that support and protect them during rising and falling tides. The arching roots are covered with **lenticels** (openings for gas exchange). Black mangroves have upright, pencil-shaped growths called **pneumatophores** that extend up from the roots and out of the water to help with aeration.

Mangrove Lenticels

Mangrove Pneumatophores

Mangrove swamps are the most threatened type of wetland. The trees are harvested for wood and cleared for development, and the habitats are used for waste dumps. Mangroves are also damaged by hurricanes on tropical islands and along the coast of Florida. Mangroves must cope with toxic soil **sulfides**—sulfur-containing chemicals that accumulate in coastal soil when oxygen is lacking. And crabs are serious predators of mangrove seeds and other plant parts.

Mangrove Prop Roots

Marsh Grasses: Common Reed (*Phragmites australis*), Saltmarsh Cordgrass (*Spartina alterniflora*), Salt Meadow Cordgrass (*Spartina patens*), Big Cordgrass (*Spartina cynosuroides*), Wild Rice (*Zizania aquatica*)

Common reed is the most widely distributed flowering plant. Especially common in alkaline habitats, it tolerates brackish water and occasional seawater flooding along the coast. In many North American wetlands, this grass is considered to be one of the most noxious invasive species. However, in areas of Europe, the species is declining, and wetland managers are working to restore populations.

Common reed is used for making brooms, baskets, mats, pen tips, paper, and roof thatching. In Australian Aboriginal cultures, these plants were used to make spears for hunting game.

Phragmites is the main wetland plant used in water treatment. Household wastewater is piped to underground tanks where solid waste settles to the bottom. Water from the top of the tank pours slowly into man-made wetlands of common reed and other Waders. Waders as well as the bacteria that live on the roots and dead leaves of the Waders absorb and remove nutrients from the wastewater. After passing through this wetland system, water is considered safe for irrigating crops or for discharging into lakes, rivers, or streams.

Phragmites Water Treatment

Saltmarsh cordgrass is one of the most important grasses in coastal wetlands. Marshlands dominated by this species offer nurseries for young fish and other animals. Salt meadow cordgrass is native to the Atlantic coast of the Americas and forms meadows inland with saltmarsh cordgrass. Big cordgrass forms dense stands in brackish and **intermediate marshes**. Big cordgrass ranges from New York and Massachusetts south to South Florida and west through Texas.

Native Americans harvest wild rice, a Wader whose grains are longer than regular rice and have black seed coats. This tall grass grows along muddy shores and in shallow brackish and fresh waters. It can be found from southern Maine to Florida, Minnesota, Indiana, Nebraska, and Texas.

Spartina Salt Marsh

Spotted Water Hemlock or Cowbane (*Cicuta maculata*), Bulbous Water Hemlock (*Cicuta bulbifera*), Poison Hemlock (*Conium maculatum*)

The hemlocks are members of the carrot family. Plants of both spotted and bulbous water hemlock produce alkaloids and other toxins that are highly poisonous to grazing animals and humans. Spotted water hemlock is considered the most toxic plant species in North America. The white tuberous roots of this species can be mistaken for parsnips. The common name, cowbane, refers to its toxicity to cows, who can die within 15 minutes after eating it. *Cicuta* species occur throughout North America.

In ancient Greece, a tea made of poison hemlock was sometimes given to condemned prisoners. It was the poison used to kill Socrates, the famous Greek philosopher and teacher, in 399 BC. Poison hemlock is native to Europe, West Asia, and North Africa, and it has been introduced into the United States.

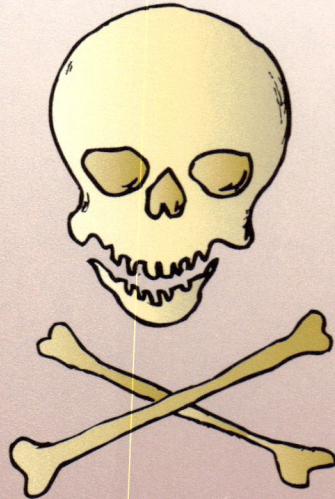

Poison Hemlock (*Conium maculatum*)

Species with Spectacular Flowers: Swamp Milkweed (*Asclepias incarnata*), Marsh Mallow (*Hibiscus moscheutos* subspecies *palustris*), Cardinal Flower (*Lobelia cardinalis*), Pickerel Weed (*Pontederia cordata*)

Swamp milkweed has breath-taking, bright rose-purple flowers clustered at the top of the plant. Butterflies and other **lepidopterans** eagerly visit the flowers. Monarch butterfly caterpillars accumulate alkaloids and glycosides from feeding on milkweed. These toxins render them distasteful and poisonous to birds and other insects. Swamp milkweed can be found in swamps, marshes, and other wet places from the northeastern United States southward into Florida. It also grows in New Mexico and Utah and in eastern Canada.

Swamp Milkweed (*Asclepias incarnata*)

Cardinal flower has many brilliant red flowers dispersed along a tall stalk. When these plants are submerged in aquaria they produce much thicker leaves and do not flower. Natural habitats include swamps, bogs, and banks of bayous, rivers, and streams from Canada, south to Florida, and west to eastern Texas.

Marsh mallows have very large flowers with creamy white petals and red centers. These shrubby marsh plants can grow to almost three meters (10 feet) tall. They are found in salt, brackish, and freshwater marshes from eastern Massachusetts, westward into the Midwest, and southward into the southeastern United States. Marsh mallows also grow in eastern Canada.

Cardinal Flower (*Lobelia cardinalis*)

Marsh Mallow (*Hibiscus moscheutos ssp. palustris*)

Pickerel weed has startling blue flowers densely clustered on a long spike. The leaf blades are either heart-shaped or lance-shaped on long petioles. Pickerel weed grows in marshes, streams, ditches, and the shallow water of ponds and lakes. Pickerel weed can be found from the northeastern United States, westward to Minnesota, and south into Florida and Texas. It also grows in eastern Canada, Central and South America, and the West Indies.

Pickerel Weed (*Pontederia cordata*)

Carnivorous Waders: Sundew (*Drosera* species), Butterwort (*Pinguicula* species), Venus Fly Trap (*Dionaea muscipula*), Pitcher Plant (*Sarracenia* species)

Carnivorous plants trap animal prey. Often growing in habitats that are low in nitrogen and phosphorus, digested prey supply carnivorous plants with these essential nutrients.

Carnivorous plants tend to grow in habitats that are maintained by fire. For example, pitcher plants, sundews, and butterworts are common in coastal Gulf state pinewoods, which experience frequent natural fires. Venus fly trap, which grows in wet savannahs with scattered pines and peat lands, is dependent on fire to reduce competition with other plants. The serious decline in wild Venus fly trap populations is probably due to the exclusion of fire from natural areas.

Sundews are **insectivorous**—that is, they consume insects. The stalked glands on sundew leaves appear to be tipped with a glistening drop of dew. These sticky, tentacle-like glands trap insects. The glands bend toward the struggling insect and move it to the center of the leaf where digestive glands are more numerous. Sundews grow in acidic soils that are constantly moist or wet. Their poorly developed roots are mostly useless for absorbing nutrients and serve only to anchor the plant and absorb water. Sundews live in bogs, pond margins, pine flatwoods, and damp sands. There are over 180 species distributed throughout the world, 50% of which occur in Australia.

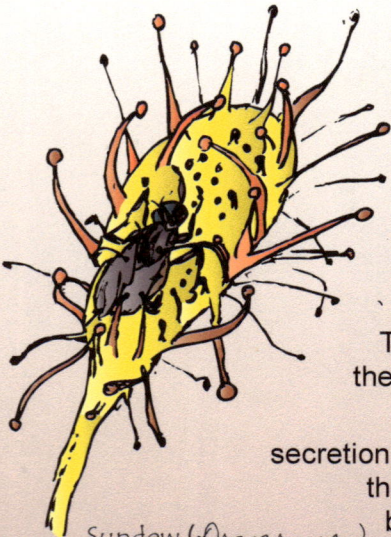

Sundew (*Drosera sp.*)

Stalked glands on butterwort leaves exude a sticky secretion that traps insects. As the insect prey struggles, the leaves curl and close, and stalkless glands on butterwort leaf surfaces secrete enzymes to digest the prey. Butterworts are often cultivated for their attractive flowers. The blue, violet, or white flowers with greenish, yellow, or reddish nectar guides are held high above the rosette of leaves.

Butterworts are in the same plant family as bladderworts. Only nine of the approximately 80 species of butterworts are native to North America. Most species occur in Central and South America. They require wet or moist habitats, growing in poor alkaline soils, on limestone rocks, in acidic and sandy soils, and in bogs.

Venus fly traps have modified leaf traps that fold in half with stiff, pointed bristles along their edges. Two to four trigger hairs in the center of each leaf-half cause the two halves of the leaf to snap shut when touched by prey. Snap-traps become inactive after trapping a few insects. Venus fly trap is in danger of extinction, with fewer than 35,000 plants in existence at last count in 1996. It is found in southeastern North Carolina, along the east coast of South Carolina, in two counties in the panhandle of Florida, and in New Jersey.

Pitcher plant leaves are tall, colorful, and funnel-shaped. The pitchers have a lid that shades the fluid inside and keeps it from evaporating or becoming diluted with rainwater. Insects that are attracted to the reddish veins or to the nectar will slip and fall into the liquid, which is laced with digestive enzymes. Downward pointing hairs prevent insects from escaping up the slippery walls inside the pitcher.

Venus Fly Trap (Dionaea sp.)

Pitchers harbor fly and mosquito larvae, **nematodes**, and **protozoa**, which are unharmed by the digestive enzymes. These organisms feed on prey that fall into the trap. Mosquito larvae live at the top of the liquid where oxygen is plentiful, while other organisms can live at lower levels of the fluid where oxygen is less concentrated or absent.

All eight species of *Sarracenia* grow in wet pinewoods, savannahs, and boggy areas of the Gulf Coastal states. Four species are confined to that area, while other species are also found in surrounding states, along the Atlantic coast, and in Canada, Oregon, and California.

Pitcher Plant (Sarracenia sp.)

Swamp Trees: Bald Cypress (*Taxodium distichum*),
Swamp Red Maple (*Acer rubrum*), Water Hickory
(*Carya aquatica*), Black Gum (*Nyssa biflora*), Water
Tupelo (*Nyssa aquatica*), Ogeechee (*Nyssa ogeche*),
Black Willow (*Salix nigra*), Water Locust (*Gleditsia
aquatica*), Buttonbush (*Cephalanthus occidentalis*)

The trees named above comprise a short representative list of those known
to grow in swamps. Most swamp trees have buttressing bases, which not
only help transport gases but also stabilize the tree.

Bald cypress is a cone bearing tree that is closely related to the
giant sequoias and coastal redwoods of the western United States.
Bald cypress forms pneumatophores (often referred to as "knees"),
which extend up from their roots a few centimeters to
a meter (few inches to several feet)

Bald Cypress (*Taxodium distichum*)
Cone & Leaves

above the ground. Bald
cypress grows abundantly in
swamps and along streams
from southern New Jersey
to Texas, up the Mississippi
Valley to southern Indiana
and Illinois.

Bald Cypress (*Taxodium distichum*)
Butresses & Pneumatophores

The large fruit of *Nyssa* species are important food sources for birds. Ogeechee grows in permanently wet swamps along rivers and streams of northern Florida and southern Georgia. The large, red ogeechee limes are sour but edible. Bees gather the abundant floral nectar to make "tupelo honey." The leaves of black gum turn purple in autumn, eventually becoming an intense bright scarlet before falling.

Tupelo Honey

Willows are small to large trees that are geographically widespread in wetland areas. They produce seeds covered with wooly hairs that allow the wind and water to carry them to new habitats. The hairiness of many willow leaves deters grazers and helps retain water.

Water Tupelo (*Nyssa aquatica*)

Black Willow (*Salix nigra*)

Buttonbush is a small tree that produces sweet scented flowers clustered in a spherical ball. It grows in temperate and tropical America. Water locust is a distinctive tree in the bean family that produces brown pods, each bearing only one seed. The trunk has long branched thorns. Water hickory, which is related to upland hickory, has a flattened hickory nut.

Southern Cattail (*Typha domingensis*), Broadleaf Cattail (*Typha latifolia*), Narrow Leaf Cattail (*Typha angustifolia*)

Cattail flowers are wind-pollinated. The tall flowering stalks bear staminate flowers above the pistillate flowers. Staminate flowers produce enormous quantities of pollen. Pistillate flowers from one plant may produce hundreds of thousands of seeds. Cattails are snorkelers and can oxygenate the soil by pumping oxygen from leaves to roots and carbon dioxide from roots to leaves.

Cattails thrive in marshes and along riverbanks and ditches. Southern cattail is the tallest of the cattails and occurs in the United States through South America in freshwater and brackish marshes. Broadleaf cattail, the most widespread of the cattails, is found in temperate areas throughout the northern hemisphere. Narrow leaf cattail is nearly as widespread in the northern hemisphere and has invaded inland from brackish coastal habitats.

Cattail (*Typha sp.*)

Classifying Habitats & Species

Wetland & Aquatic Habitats: Natural habitats occur along a moisture gradient from dry uplands to wetlands and aquatic environments, and it is difficult to pinpoint where uplands end and wetlands begin. Water levels change daily with tides, seasonally with rainfall, and over longer geologic time periods of fluctuating global warming and cooling (glacial and interglacial periodicity). During years of high precipitation, wetlands increase their boundaries to cover territory previously considered uplands. In contrast, wetlands become drier as plant sediments build up through the course of ecological succession. Eventually, such former wetlands can become dry uplands occupied by corresponding upland plant and animal life.

Modern wetland plant ecologists agree that plants in wet habitats fall into the categories of Submerged, Floating, Floating-Leaved, and Emergent. In this book, we use more user-friendly terms that are easier to remember: Divers, Floaters, Floating-Leaf Plants, and Waders.

Some plant ecologists classify wetland habitats as those that are in or on the water or that have frequently flooded (and therefore anaerobic) soils. They consider these habitats *wetlands*, regardless of whether they are inundated with water daily (e.g., tidal marshes), are seasonally wet (e.g., freshwater marshes, margins of streams, and rivers during spring snow melts), or have standing water (e.g., ponds, lakes, rivers, and streams). Other scientists distinguish between *wetlands* (where Waders abound) and *aquatic habitats* (where permanent waters exist, and Divers, Floaters, and Floating-Leaf Plants thrive).

Classifying Species: Species described in this book are identified by their scientific and common names. When taxonomists describe new species, they give them scientific names, which consist of two parts, or a binomial. The word *binomial* comes from Latin; *bi* means *two*, and *nom* means *name*. The first part is the genus, which is capitalized. The second part is the specific epithet, which is lower case. For example, the scientific name for water tupelo is *Nyssa aquatica*; *Nyssa* is the genus, and *aquatica* is the specific epithet. A genus is like a family surname and may contain one to many species. For example, ogeechee is related to water tupelo, and its scientific name is *Nyssa ogeche*.The binomial system of naming has been in use since its invention by Carl Linnaeus in 1753. A set of rules for naming new species insures that each species has only one binomial scientific name that belongs to that species only.

In contrast to scientific names, *common names* are not as tightly regulated. Common names often derive from what local people call the species. The same plant can have many common names, and those common names are not always unique to an individual species. For example, *Phragmites australis* is the scientific name for a tall grass that grows in marshes almost worldwide. In the brackish coastal areas of Louisiana where plants of this species are used for building duck blinds, the people call it Roseau cane. Throughout Europe where the stems are used to thatch roofs, it is called common reed, Norfolk reed, and water reed. In the Philippines, the local name is tambo. In December, stalks and flowers of *Phragmites* are cut and bundled into brooms or walis. The brooms made of tambo are referred to as walis tambo.

Over 250,000 species of flowering plants have been discovered and assigned scientific names. Scientists estimate that 30,000-70,000 flowering plant species remain undiscovered and unnamed. For plants, there has been no attempt to standardize common names. Such a task, in addition to scientific naming of perhaps 300,000 or more species, would be overwhelming and would create confusion when changing or restricting common names that have been used for centuries.

Preservation of Wetland Habitats

Wetland Habitat Loss: Today, wetlands cover approximately 9 million square kilometers (3.5 million square miles)—that is, 6% of the earth's land area. In the continental U.S., 5% of the land is wetland. Loss of wetland habitats has been extensive worldwide, primarily due to land use conversion to agriculture; other losses have resulted from chemical pollutants such as fertilizers and herbicides, invasion of exotic species, lumbering in forested swamplands, exclusion of fires in certain habitats, and altering water flow patterns. Wetland losses are especially severe in more populated areas.

Nearly 40% of wetland habitats in the U.S. occur in Louisiana, where the rate of wetland loss is astounding. Between 1932 and 2010, Louisiana lost 25% of its wetlands. Early European settlers built levees along the Mississippi River to protect towns from floods, and when floodwaters topped those levees, they built higher ones. Levees channelized the Mississippi River and robbed wetlands along the river as well as the downstream coastal marshes of the flow of sediments. Without a sediment supply, coastal marshlands eroded; plants died, leaving behind open water where marshes existed before. Concurrent with levee building, settlers also began to cut and remove the enormous, virgin, bald cypress trees from the swamps.

Hurricanes also threaten wetlands. For example, the major hurricanes that struck coastal Louisiana (Andrew in 1992, Katrina and Rita in 2005, Gustav and Ike in 2008), caused an average annual loss of 43 square kilometers (17 square miles) of wetlands.

Intensive oil and gas industry development in Louisiana damages coastal wetlands. To gain access to well sites, companies dig canals. Dredged sediments pile along canal banks, smothering plant life and impeding the flow of water and sediments that sustain coastal marshes. In addition, extraction of oil and gas from underground causes the marshes to subside, or sink, subsequently being replaced by open water.

In 2010, the British Petroleum Deepwater Horizon oil spill killed marsh plants growing 4.6-9 meters (15-30 feet) from the shoreline. Because the roots of heavily oiled grasses died, they could no longer hold onto sediments, and erosion from wave action along the marsh edges more than doubled.

Sea level rises affect coastal wetlands worldwide. Over time, elevations of marshes rise naturally through sedimentation and vegetation decay. However, if sea levels rise too rapidly, natural sedimentation can be outpaced, and coastal wetlands retreat inland, if possible, or disappear. The U.S. Environmental Protection Agency predicted that during the next century, sea levels will rise 15-34 centimeters (6-13 inches), which would inundate coastal wetlands. Over the long-term, sea levels rise during interglacial periods, like the current one, but the current pace of rise is higher than in previous interglacial periods.

As a result of global warming, the earth may experience hotter, wetter, and wilder weather. Extreme spikes in temperature, shifting rainfall patterns, more intense storms, destructive storm surges in coastal areas, and flooding will most likely severely decrease wetland area. Some habitats will retract, while others expand. Stressed ecosystems tend to invite invasion of exotic species. Slow growing plants, such as swamp trees, will probably be at the greatest risk from future drought conditions. In contrast, smaller plants that reproduce rapidly, spread easily from place to place, reproduce sexually, and thus have greater genetic diversity, will have the greatest advantage.

Cronk and Finnessy (2001) listed over 120 species of endangered and threatened wetland plant species, many of which occur in endangered habitats. They asserted that the proportion of wetland species that are threatened or endangered is significantly higher than that of terrestrial species. The renowned biologist, E. O. Wilson, predicted that we are in the midst of the sixth mass extinction. He estimated that the earth is losing 27,000 species per year—that is, 74 per day, or three per hour. No doubt, given the continued stresses of climate change and lack of sufficient time to adapt to changing environments, many wetland plants species will be lost during this period of mass extinction.

Protecting Wetland Habitats: Humans are increasingly recognizing that healthy wetland habitats benefit the health and welfare of humans, wildlife, fishing industries, aquatic recreation, and water quality. Therefore, many humans are endeavoring to slow the pace of wetland habitat loss. Wetland habitats depend on healthy native wetland plants, which require proper gas exchange, light, support, and reproduction. Legal definitions and delineations of wetland habitats have been established in many countries, including the , to protect and regulate the habitat use.

The shorelines, or boundaries, between uplands and navigable waters were originally demarcated by pre-statehood surveyors from the U.S. General Land Office who mapped "meander lines," (i.e., ordinary high water lines). The surveyors used straight-line methods that lacked correspondence with the natural, gentle curves of land that surround water bodies. However, at the time, the goal of the surveying was to aid in navigation and not to protect wetlands. The U.S. General Land Office "patented," or granted title of, navigable bodies of water (i.e., lakes 10 hectares (25 acres) or larger and streams 60 meters (197 feet) or more in width) to the states. Non-navigable wetland habitats such as swamps, marshes, ponds, and smaller streams and lakes were not recognized, and therefore, they were relegated to the category of uplands that could be drained and used for agriculture and other types of development.

In 1972, with growing public awareness of the importance of protecting wetland habitats, the U.S. Congress approved the Clean Water Act, which established government regulation of wetlands by the Environmental Protection Agency, the U.S. Army Corp of Engineers, and other agencies. This act and later amendments insure the regulation of all waters with "significant nexus" (i.e., connections with navigable waters, defined as "relatively permanent, standing or continuously flowing bodies of water"). In addition, under the Clean Water Act, any dredging or filling that affects the bottom elevation of wetlands requires a permit issued by the U.S. Army Corp of Engineers.

Despite government regulations, controversy over government control of private property and limited fiscal resources necessary to enforce the regulations hinder wetland protection.

GLOSSARY

AERENCHYMA: a tissue in plants with large air spaces between cells.

AEROBIC RESPIRATION: breaking down of food in respiration requiring oxygen and yielding 36 ATP molecules.

ALGAE: a large, diverse group of photosynthetic organisms that range in complexity from microscopic single cells, such as *Chlorella*, to large multicellular bodies, such as kelp.

ALKALOIDS: plant compounds that are often poisonous.

ANAEROBIC RESPIRATION: breaking down of food in respiration in the absence of oxygen that yields only two ATP molecules.

ANTHER: the pollen containing part of the stamen of the flower.

AQUARIUM: a water filled tank for keeping live plants and animals, usually with glass sides and lighting.

ASEXUAL REPRODUCTION: producing new individuals without fertilization—that is, without the union of sperm and egg.

ATP: adenosine triphosphate, an energy-carrying molecule found in all living organisms.

BRACKISH: freshwater mixed with saltwater but not as salty as seawater.

BRYOPHYTES: a collective term for mosses, liverworts, and hornworts, a primitive group of small plants with no true conducting tissues.

BUTTRESSES: swellings in the trunks of trees in response to flooding.

CARBOHYDRATE: a collective term for organic compounds made up of carbon, hydrogen, and oxygen. Sucrose, table sugar, is a type of carbohydrate made up of two simple sugars. Starch and cellulose are examples of complex carbohydrates made up of several hundred glucose molecules linked together.

CARNIVOROUS: meat-eating.

CELLULOSE: a complex carbohydrate that forms the walls of plant cells.

CENTRAL VACUOLE: a large membrane-bound sack found in plant cells.

CLONING: producing new individuals that are genetically identical to the parent organism.

CONE: the reproductive body of conifers, such as pine, with overlapping, often woody scales.

CROSS-POLLINATION: delivery of pollen to stigmas or female cones of another individual, in contrast to self-pollination.

CUTICLE: a waxy layer on the surface of epidermal plant cells.

CYANOBACTERIA: a group of bacteria that have a blue-green colored pigment. *Cyano* means dark blue. These bacteria can make their own food through photosynthesis. Many are nitrogen fixers.

DIFFUSION: the movement of substances from regions of higher to lower concentrations.

ECOLOGISTS: scientists who study the relationships among organisms in their habitats.

ELECTRON: the negative particle of atoms that carries electricity.

EMBRYO: an organism in the early stages of growth.

EMERGENT: a term meaning *above* and in reference to wetland plants, meaning *above water*.

ENDOSPERM: a nutritive tissue in seeds, which provides food for the developing embryo upon germination.

EPIDERMIS: a tissue of plants and animals forming a protective outer layer.

FRESHWATER: water containing little or no salt.

FROND: a leaf or leaf-like plant part, which may refer to both leaf and stem together.

GENUS: the first part of a scientific name. For example, in the scientific name for human beings, *Homo sapiens, Homo* is the genus, and *sapiens* is the specific epithet.

GERMINATE: to begin to grow; as in the process by which seeds open and grow into new plants or pollen grows a tube into the style.

GLUCOSE: a type of carbohydrate known as a simple sugar that is made up of six carbon atoms, twelve hydrogen atoms, and six oxygen atoms.

GUARD CELLS: the two cells surrounding the opening of a stomate.

HABITAT: the place where plants, animals, and other organisms live.

INSECTIVOROUS: insect-eating.

INTERMEDIATE MARSHES: habitats that are intermediate between brackish and freshwater in saltiness.

INVASIVE: organisms that establish and spread rapidly in habitats in which they are not native. Invasive species most often are introduced from distant continents or areas of the world.

LENTICELS: openings in stems and roots for gas exchange.

LEPIDOPTERANS: butterflies and their relatives, the skippers and moths.

LIGNIN: the substance that makes wood hard and strong.

MARSH: a treeless wetland covered by Wader plants.

METHANE: CH_3 gas.

MNEMONIC: a method of remembering.

MYCORRHIZAL FUNGUS: a fungus symbiotic with plant roots.

NADPH: nicotinamide adenine dinucleotide phosphate, an energy rich molecule produced by the light reaction of photosynthesis.

NECTAR: a sweet liquid in flowers that bees, hummingbirds, and other animals use for food.

NEMATODE: a small worm-like organism with a tubular digestive system that has openings at both ends.

NITRATE: one chemical form of nitrogen combined with oxygen; an important fertilizer for plants and needed to make proteins and DNA.

NITROGEN FIXATION: the process of converting nitrogen gas into ammonia by bacteria.

ORGANELLES: small bodies within cells.

OXIDIZED: combined with oxygen.

PEDUNCLE: stalk bearing a flower or flowers.

PETALS: modified leaves that surround the reproductive flower parts.

PERFECT FLOWERS: flowers with both stamens and pistils.

PETIOLE: the stalk of a leaf that attaches the leaf blade to the stem.

PHLOEM: the living tissue of vascular plants that transports organic nutrients, especially sugar products of photosynthesis.

PHOSPHATE: a chemical compound of phosphorus and oxygen; an important fertilizer for plants and needed for making ATP and other molecules.

PHOTOSYNTHESIS: a process that converts light energy, carbon dioxide, and water into glucose.

PHOTOSYNTHETIC: organisms that make their own food using light.

PHYSIOLOGISTS: scientists who study how organisms function.

PISTIL: the female reproductive part of the flower.

PISTILLATE FLOWERS: flowers that have pistils but no stamens.

PNEUMATOPHORES: structures for aeration that grow from roots.

POLLEN: the male part of the plant that contains sperm to fertilize eggs.

POLLINATION: delivery of pollen from the anther to the stigma of the flower in flowering plants, or from the male cone to the female cone in conifers.

POLLINATOR: an animal that transfers pollen within or among flowers.

POLLUTED: unclean, dirty.

PRISM: a transparent object that splits sunlight into its spectrum of colors.

PROPAGULES: asexual plant parts that grow into new plants.

PROTOZOA: single-celled animal-like organisms in the Kingdom Protista.

REDUCED SOILS: soils with low oxygen content.

RHIZOMES: horizontal stems that either burrow into or grow on top of the ground.

ROOT HAIRS: single-cell structures found at the tips of roots and which absorb water and dissolved minerals.

ROOT NODULES: swellings in roots that contain nitrogen-fixing bacteria.

ROSETTE: a plant form in which leaves are arranged in a circular pattern, like the petals of a rose.

RUNNERS: stem of a plant that grows horizontally along the ground.

SALTWATER: salty water as in the oceans and seas.

SELF-POLLINATION: when pollen is transferred to a pistil within the same flower or a flower on the same plant.

SPORE: an asexual reproductive structure capable of growing into a new individual.

STAMEN: the part of the flower consisting of an anther, where pollen is made, and a filament stalk.

STAMINATE FLOWERS: flowers that have stamens but no pistil.

STIGMA: the top of the pistil where pollen germinates.

STOLONS: stems that grow from the parent and produce new plants.

STOMATES: small openings on the surface of leaves and stems that allow gases to pass in and out.

STYLE: the neck-like part of the pistil between the stigma and the ovary.

SULFIDES: chemicals containing sulfur, such as hydrogen sulfide (H_2S), which are common in reduced soil and are toxic to plants.

SWAMP: a body of water that can dry periodically; a type of wetland with trees.

SYMBIOTIC: a mutually beneficial relationship between organisms.

TAXONOMISTS: scientists who classify and name organisms.

TISSUES: groups of cells that perform specialized functions.

TOXIC: poisonous.

TRANSPIRATION: a type of evaporation from plants through stomates; a process important in helping to pull water from the roots to the tops of plants.

TUBER: any potato-like underground stem.

TURGID: full and tight, as cells become when they take in water.

TURION: an overwintering bud produced by water plants for asexual reproduction.

UPLAND PLANTS: plants that grow best in elevated areas without standing water either seasonally or for sustained periods.

VASCULAR PLANTS: organisms in the plant kingdom with specialized xylem and phloem cells for transporting water and dissolved minerals and sugar.

WAVELENGTH: a measure of electric and magnetic properties of light, sound, and water. Light is emitted in packets of waves with peaks and valleys. Wavelength is the distance between peaks.

WETLAND: a seasonally or permanently flooded area.

XYLEM: tubes connected end-to-end like pipes inside stems of plants that are responsible for transporting water and dissolved minerals.

REFERENCES

BIOLOGY AND ECOLOGY OF AQUATIC AND WETLAND PLANTS

Cronk, Julie K. and M. Siobhan Fennessy. 2001. *Wetland Plants: Biology and Ecology*. Boca Raton, FL: CRC/Lewis Publishers.

Dacey, John W. H. 1981. Pressurized Ventilation in the Yellow Waterlily. *Ecology* 62: 1137-1147.

Kalman, Bobbie. 2008. *Photosynthesis: Changing Sunlight into Food (Nature's Change)*. New York: Crabtree Publishing Company.

McPherson, Stewart. 2008. *The Glistening Carnivores: The Sticky-Leaved Insect-eating Plants*. Dorset, England: Redfern Natural History Productions.

Meyers, Dewey G. and J. R. Strickler. 1979. Capture Enhancement in a Carnivorous Aquatic Plant: Function of Antennae and Bristles in *Utricularia vulgaris*. *Science* 203: 1022-1024.

Niering, William. 1985. *Wetlands*. New York: A. A. Knopf.

IDENTIFICATION MANUALS FOR REGIONS IN THE UNITED STATES

Chadde, Steve W. 2012. *A Great Lakes Wetland Flora: A Complete Guide to the Wetland and Aquatic Plants of the Midwest*. Charleston, SC: Createspace Independent Publishing Platform.

Correll, Donovan Stewart. 1975. *Aquatic and Wetland Plants of Southwestern United States*. Palo Alto, CA: Stanford University Press.

Crow, Garrett E. and C. Barre Hellquist. 2006. *Aquatic and Wetland Plants of Northeastern North America: Revised and Enlarged Edition of Norman C. Fassett's A Manual of Aquatic Plants*. 2 vols. Madison, WI: Univ. of Wisconsin Press.

Godfrey, Robert K. and Jean W. Wooten. 1979. *Aquatic and Wetland Plants of Southeastern United States: Monocotyledons*. Athens, GA: Univ. of Georgia Press.

———. 1981. *Aquatic and Wetland Plants of Southeastern United States: Dicotyledons*. Athens, GA: Univ. of Georgia Press.

Lahring, Heinjo. 2003. *Water and Wetland Plants of the Prairie Provinces: A Field Guide for Alberta, Saskatchewan, Manitoba and the Northern United States*. Regina, Saskatchewan, Canada: Canadian Plains Research Center.

Mason, Herbert L. 1969. *A Flora of the Marshes of California*. Berkeley, CA: Univ. of California Press.

Newcomb, Lawrence. 1989. *Newcomb's Wildflower Guide*. New York: Little, Brown and Company.

Steward, Albert N., La Rea J. Denis, and Helen M. Gilkey. 1963. *Aquatic Plants of the Pacific Northwest with Vegetative Keys*. Corvalis, OR: Oregon State Univ. Press.

GROWING AQUATIC PLANTS IN AQUARIA

Hiscock, Peter. 2005. *Aquarium Plants (Mini Encyclopedia Series for Aquarium Hobbyists)*. Hauppauge, NY: Barron's Educational Series, Inc.

Sweeney, Mary E. 2001. *The Guide to Owning Aquarium Plants*. Neptune City, NJ: TFH Publications, Inc.

Sweeney, Mary E. George Farmer, Neil Hepworth, Aaron Norman, Jeff Ucciardo. 2008. *The 101 Best Aquarium Plants: How to Choose Hardy, Vibrant, Eye-Catching Species That Will Thrive in Your Home Aquarium*. Neptune City, NJ: TFH Publications, Inc.

PRESERVATION OF WETLAND HABITATS

Couvillion, Brady R., John A. Barras, Gregory D. Steyer, William Sleavin, Michelle Fischer, Holly Beck, Nadine Trahan, Brad Griffin, and David Heckman. 2011. Land Area Change in Coastal Louisiana from 1932 to 2010. U.S. Geological Survey. http://pubs.usgs.gov/sim/3164/downloads/SIM3164_Pamphlet.pdf.

Executive Order no. 11990. 1977. Protections of Wetlands. U.S. Environmental Protection Agency. http://water.epa.gov/lawsregs/guidance/wetlands/eo11990.cfm.

Higgins, Jerome S. 1887. Subdivisions of Public Lands: Described and Illustrated with Diagrams and Maps. St. Louis, MO: Higgins and Co.

Kemp, Andrew C., Benjamin P. Horton, Jeffery P. Donnelly, Michael E. Mann, Martin Vermeer, and Stefan Rahmstorf. 2011. Climate Related Sea-Level Variations Over the Past Two Millennia. *Proceedings of the National Academy of Sciences* 108 (27): 11017-11022.

Kopp, Robert E., Frederick J. Simon, Jerry X. Mitrovica, Adam C. Maloof, and Michael Oppenheimer. 2009. Probabilistic Assessment of Sea Level During the Last Interglacial Stage. *Nature* 462: 863-867.

Mitsch, W. J. and J. G. Gosselink. 2000. *Wetlands*. 3rd ed. New York: John Wiley & Sons.

Poore, Richard Z., Richard S. Williams, Jr., and Christopher Tracey. 2011. Sea Level and Climate. U.S. Geological Survey. http://pubs.usgs.gov/fs/fs2-00.

Silliman, Brian R., Johan van de Koppel, Michael W. McCoy, Jessica Diller, Gabriel N. Kasozi, Kamala Earl, Peter N. Adams, and Andrew R. Zimmerman. 2012. Degradation and resilience in Louisiana salt marshes after the BP–Deepwater Horizon oil spill. *Proceedings of the National Academy of Sciences*. 109 (28): 11234-11239.

Wilson E. O. 1999. *The Diversity of Life*. 2nd ed. New York: W.W. Norton & Co, Ltd.

INDEX

www.ingramcontent.com/pod-product-compliance
Lightning Source LLC
Chambersburg PA
CBHW060809270326
41928CB00002B/32